# Collaboration and Knowledge Sharing in Armyworm Management

Mogana S. Flomo, Jr.

Published by MGI Inc., 2023.

While every precaution has been taken in the preparation of this book, the publisher assumes no responsibility for errors or omissions, or for damages resulting from the use of the information contained herein.

COLLABORATION AND KNOWLEDGE SHARING IN ARMYWORM MANAGEMENT

**First edition. June 15, 2023.**



ISBN: 979-8223574064

Written by Mogana S. Flomo, Jr..

# Dedication

This book is dedicated to all farmers in Liberia who are tirelessly combating the challenges posed by Armyworm infestations. Your resilience, dedication, and commitment to ensuring food security and agricultural sustainability inspire us all. May this book serve as a valuable resource, providing you with insights, strategies, and knowledge to effectively manage Armyworms and protect your livelihoods.

We also extend our heartfelt appreciation to all stakeholders in both the public and private sectors who are actively involved in supporting farmers and agricultural development. Your collaboration, expertise, and resources are crucial in mitigating the impact of Armyworm outbreaks and fostering a sustainable agricultural landscape in Liberia.

Together, let us continue our collective efforts to empower farmers, promote innovative solutions, and build a resilient farming community capable of overcoming the challenges posed by Armyworms. Your unwavering dedication is the backbone of Liberia's agricultural sector, and we are grateful for your unwavering commitment.

*Preface*

Welcome to "Collaboration and Knowledge Sharing in Armyworm Management: Strengthening Stakeholder Engagement for Sustainable Agriculture." This book is a culmination of extensive research, practical experiences, and a shared commitment to addressing the challenges posed by armyworm infestations and their socioeconomic impacts on farming communities.

Armyworms have emerged as a formidable threat to agricultural productivity, affecting crops such as maize, rice, sorghum, and vegetables in various regions around the world. These pests not only cause substantial crop damage but also pose significant economic and food security risks for farmers and communities reliant on agriculture. Effectively managing armyworm infestations requires a collective effort, involving stakeholders from government agencies, NGOs, research institutions, and farmers.

The key focus of this book is to explore the power of collaboration and knowledge sharing in armyworm management. We firmly believe that by bringing stakeholders together, fostering cooperation, and sharing knowledge, we can develop innovative and sustainable solutions to combat armyworm infestations and mitigate their impacts.

Throughout the book, we delve into the importance of collaboration as a foundational element in addressing the armyworm challenge. We provide practical guidance on establishing effective partnerships, creating collaborative frameworks, and leveraging digital tools and platforms for enhanced knowledge sharing. We understand that collaboration is not just about working together but also about building trust, understanding diverse perspectives, and creating synergies among stakeholders.

Additionally, this book highlights the socioeconomic implications of armyworm infestations on farming communities. We discuss the role of governments in implementing national armyworm control programs, providing subsidies for inputs, and supporting research and development initiatives. We emphasize the significance of extension services and farmer training in disseminating information and building capacity at the grassroots level.

Furthermore, we present case studies that showcase successful collaboration and knowledge sharing initiatives in armyworm management. These examples provide practical insights, lessons learned, and inspiration for stakeholders engaged in similar efforts. We believe that sharing these experiences will promote the adoption of best practices and contribute to the collective knowledge base in this field.

To facilitate a common understanding, we have included a glossary that defines key terms and concepts related to collaboration, knowledge sharing, and armyworm management. We hope this will serve as a helpful reference for readers as they navigate the book.

Lastly, we provide an extensive list of additional readings and references, allowing readers to explore specific topics in greater depth. We encourage you to utilize these resources to expand your knowledge and stay informed about the latest developments in armyworm management and sustainable agriculture.

We extend our heartfelt appreciation to the researchers, practitioners, and stakeholders who have contributed to this book. Your expertise, insights, and dedication to collaboration and knowledge sharing have been instrumental in shaping its content.

We sincerely hope that "Collaboration and Knowledge Sharing in Armyworm Management" serves as a valuable resource for all stakeholders involved in combating armyworm infestations and promoting sustainable agriculture. Together, let us build a future where resilient collaborations and shared knowledge lead to effective armyworm management, thriving farming communities, and a sustainable food system.

*Mogana S. Flomo, Jr.,.*

*Description:*

This book offers a comprehensive guide to collaboration and knowledge sharing in the context of armyworm management, with a focus on fostering effective stakeholder engagement for sustainable agricultural practices. It explores the challenges posed by armyworm infestations and their socioeconomic impacts on farming communities. Drawing on case studies, research findings, and practical insights, this book provides valuable strategies and tools for stakeholders involved in armyworm management, including government agencies, NGOs, researchers, and farmers.

The book begins by highlighting the importance of collaboration and knowledge sharing in addressing the threat of armyworm infestations. It delves into the various approaches and techniques used to facilitate collaboration among stakeholders, such as online collaboration platforms, content management systems, and knowledge management software. It also presents a template collaboration agreement that stakeholders can use as a reference when forming partnerships or joint initiatives.

Furthermore, the book explores the impact of armyworms on farming communities and the socioeconomic factors at play. It discusses the role of governments in implementing national armyworm control programs and providing subsidies for pesticides and inputs. The importance of extension services and farmer training in disseminating information and building capacity is emphasized.

The book also covers research and development support for armyworm control, including funding initiatives and the development of innovative control strategies. It highlights the significance of regional collaboration and cooperation among countries facing armyworm infestations for collective action and resource sharing.

To enhance readers' understanding, the book includes a glossary section that defines key terms and concepts related to collaboration, knowledge sharing, and armyworm management. Additionally, it provides an extensive list of additional readings and references for those interested in delving deeper into the topic.

Overall, this book serves as a comprehensive resource for stakeholders seeking to improve collaboration and knowledge sharing in armyworm management. It offers practical guidance, real-world examples, and valuable insights to foster sustainable agriculture practices and mitigate the impact of armyworm infestations on farming communities.

# 1 Introduction to Armyworms

## 1.1 Background on armyworms

Armyworms are a group of insect pests that belong to the order Lepidoptera and the family *Noctuidae*. They are known for their destructive feeding habits and their ability to cause significant damage to various crops. Armyworm species are widely distributed around the world and are particularly prevalent in tropical and subtropical regions.

The life cycle of armyworms typically consists of four stages: egg, larva (*caterpillar*), pupa, and adult. The adult armyworm moths lay their eggs on host plants, which can include a wide range of crops such as maize, rice, wheat, sorghum, and vegetables. The eggs hatch into larvae, commonly referred to as armyworm caterpillars, which feed voraciously on plant foliage.

As the caterpillars grow, they go through several instar stages, during which they molt and increase in size. Armyworms are known for their behavior of moving in large groups or "armies," which gives them their name. They exhibit mass migration patterns, traveling in search of new food sources, hence posing a threat to crops over large areas.

Once the caterpillars have completed their feeding stage, they enter the pupal stage, during which they undergo metamorphosis. After the pupal stage, adult moths emerge, mate, and begin the cycle again by laying eggs on suitable host plants.

Armyworm infestations can have devastating effects on agriculture, leading to significant crop damage and yield losses. The feeding activity of armyworm caterpillars can result in defoliation, stem cutting, and even complete destruction of crops, depending on the severity of the infestation. The economic impact of armyworms is substantial, with estimated annual losses in crop production reaching billions of dollars globally.

Several species of armyworms are of agricultural significance, including the fall armyworm (*Spodoptera frugiperda*), African armyworm (*Spodoptera exempta*), and the common armyworm (*Mythimna unipuncta*). These species have different geographic distributions but share similar feeding habits and destructive potential.

The susceptibility of crops to armyworm infestations depends on various factors, including the crop type, stage of growth, weather conditions, and natural enemy populations. Climate variability, such as changes in temperature and rainfall patterns, can influence the frequency and intensity of armyworm outbreaks.

Numerous studies have focused on the biology, behavior, and management of armyworms. Researchers have investigated the factors influencing armyworm outbreaks, including climate change, land use practices, and the presence of natural enemies. Understanding the life cycle and behavior of armyworms is crucial for developing effective management strategies and mitigating their impact on agricultural systems.

In summary, armyworms are a group of insect pests that pose a significant threat to agriculture. Their destructive feeding habits and ability to cause crop damage have serious socioeconomic implications for farming communities. Extensive research has been conducted to understand their biology and develop strategies for their management.

## 1.2 Significance of armyworm impact on farming

ARMYWORM INFESTATIONS have significant implications for farming practices, agricultural productivity, and the overall socioeconomic well-being of farming communities. The following points highlight the significance of armyworm impact on farming:

1. *Crop Losses and Yield Reduction*: Armyworms are voracious feeders that can quickly consume large portions of crops. Their feeding behavior leads to defoliation, damage to reproductive structures, and even complete destruction of plants. As a result, farmers experience substantial crop losses and reduced yields, leading to economic hardships and food insecurity (Sisay et al., 2018; Gahukar, 2017).

2. *Economic Losses*: The economic impact of armyworm infestations is substantial. Farmers not only bear the cost of lost crop production but also face increased expenses for pest control measures and the use of insecticides. The financial burden is often exacerbated by the need for repeated interventions to mitigate armyworm outbreaks, leading to decreased profitability and financial stability for farmers (Kansiime et al., 2021; Alem et al., 2018).

3. *Disruption of Livelihoods*: Farming communities heavily rely on agriculture for their livelihoods and income. Armyworm infestations threaten the stability of these communities by undermining their primary source of economic activity. The reduced income and employment opportunities resulting from crop losses can push farmers and their families into poverty and exacerbate existing socioeconomic disparities (Anderson et al., 2019; Attignon et al., 2020).

4. *Food Security Concerns*: The impact of armyworms extends beyond individual farmers, affecting the broader food security of regions and countries. As armyworms target staple crops such as maize, rice, and wheat, infestations can disrupt local and regional food supplies, leading to higher food prices and decreased availability of nutritious food. Vulnerable populations, particularly those with limited access to

alternative food sources, are most affected by these disruptions (Kumela et al., 2021; Midega et al., 2018).

5. *Agricultural Productivity and Sustainability*: Armyworm infestations undermine agricultural productivity and pose challenges to sustainable farming practices. Farmers may resort to excessive use of chemical pesticides to control the pests, leading to environmental pollution and negative impacts on beneficial insects, soil health, and biodiversity. Additionally, the need for intensive pest management practices diverts resources and attention away from implementing sustainable and environmentally friendly agricultural practices (Castañeda et al., 2021; Pachanga-Piñeda et al., 2019).

In summary, the significance of armyworm impact on farming is profound, encompassing crop losses, economic hardships, food security concerns, and challenges to sustainable agricultural practices. Addressing and mitigating the impact of armyworm infestations is crucial for ensuring the stability and well-being of farming communities and agricultural systems.

## 1.3 Importance of socioeconomic status of farming communities

THE SOCIOECONOMIC STATUS of farming communities plays a vital role in shaping their resilience, adaptability, and overall well-being in the face of armyworm infestations. Here are key points highlighting the importance of socioeconomic status in farming communities:

1. *Livelihood Dependence*: Farming communities often heavily depend on agriculture as their primary source of income and livelihood. The success or failure of agricultural activities,

including the impact of armyworm infestations, directly affects the economic well-being of these communities. Income generated from farming supports various aspects of their lives, including food security, education, healthcare, and other essential needs (Akudugu et al., 2020; Shiferaw et al., 2014).

2. *Vulnerability and Poverty*: Many farming communities, particularly smallholder farmers in developing regions, face economic vulnerabilities and are more susceptible to the adverse effects of armyworm infestations. Limited access to financial resources, lack of infrastructure, and limited technical knowledge and capacity exacerbate their vulnerability. Armyworm outbreaks can push these communities further into poverty, perpetuating a cycle of economic hardship (Andersson et al., 2021; Nkonya et al., 2016).

3. *Food Security and Nutrition*: The socioeconomic status of farming communities directly influences their access to and availability of nutritious food. Armyworm infestations can undermine food security by reducing crop yields and availability of staple crops. This not only impacts farmers' own households but also affects their ability to contribute to local and regional food supplies. Consequently, communities may face increased food prices, limited dietary diversity, and a higher risk of malnutrition (Fischer et al., 2019; Ogada et al., 2020).

4. *Education and Human Capital*: The socioeconomic status of farming communities has implications for education and human capital development. Limited financial resources resulting from armyworm-related losses may hinder access to quality education for children in farming families. The lack of educational opportunities can further perpetuate the cycle of poverty and hinder the community's ability to adapt and innovate in response to agricultural challenges (Reardon et al.,

2019; World Bank, 2020).

5.  *Community Resilience and Adaptation*: Strong socioeconomic status provides a foundation for community resilience and adaptation to armyworm infestations. Communities with higher socioeconomic resources, such as access to credit, technical assistance, and social networks, are better equipped to cope with the impacts. They can invest in improved farming practices, diversify income sources, and access information and resources necessary to mitigate and manage armyworm outbreaks effectively (Darnhofer et al., 2016; Sumberg et al., 2013).

Recognizing the importance of the socioeconomic status of farming communities is crucial in designing interventions and policies that address the specific needs and challenges they face. Supporting the economic well-being, access to resources, and capacity development of these communities can enhance their resilience and ability to cope with armyworm infestations, ultimately contributing to sustainable agriculture and rural development.

# 2 A closer look at Armyworms

## 2.1 Description and Life Cycle of Armyworms

As mentioned earlier, Armyworms are a group of moth larvae belonging to the genus *Spodoptera,* specifically the species *Spodoptera exempta* (African armyworm) and *Spodoptera frugiperda* (fall armyworm). They are named after their behavior of forming large "army-like" groups as they migrate and feed on crops.

### 2.1.1 *Physical* Description:

ARMYWORM LARVAE ARE caterpillar-like in appearance, with cylindrical bodies that can range in color from green to brown or black, depending on the species and instar (developmental stage).

They have six true legs near the head and several prolegs along the abdomen.

The head is usually dark with strong mandibles for feeding.

### 2.1.2 Life Cycle:

THE LIFE CYCLE OF ARMYWORMS consists of four main stages: egg, larva, pupa, and adult.

*a. Egg Stage*: Female moths lay eggs in clusters, usually on the undersides of leaves or other protected surfaces.

The eggs are small, spherical, and may vary in color from white to pale green or gray.

The incubation period typically lasts for several days, depending on environmental conditions.

*b. Larva (Caterpillar) Stage*: After hatching, the armyworm larvae go through several instars, during which they actively feed and grow.

The larvae exhibit gregarious behavior, moving in groups and consuming large amounts of vegetation.

They undergo a series of molts, shedding their exoskeletons to accommodate their increasing size.

The larval stage can last from a few weeks to several months, depending on environmental factors and food availability.

*c. Pupa Stage*: When the larvae reach full size, they enter the pupal stage.

The pupa is typically brown or dark in color and is enclosed within a cocoon formed by the larva.

During this stage, the insect undergoes metamorphosis, transforming into an adult moth.

*d. Adult Stage*: After the pupal stage, the adult moth emerges from the cocoon. Adult armyworm moths have a wingspan of about 3-4 centimeters. The coloration and patterns on their wings can vary among species. Adult moths are primarily nocturnal and are attracted to lights.

Figure 1: Life Cycle of Fall Armyworm. Diagram courtesy of Fall Armyworm larval identification guide DPIRD.

THE LIFE CYCLE OF ARMYWORMS is influenced by environmental conditions, particularly temperature and humidity. Under favorable conditions, they can complete multiple generations in a year, contributing to their potential for rapid population growth and widespread infestations.

Understanding the life cycle of armyworms is crucial for developing effective management strategies and timing interventions to target vulnerable stages of their development. Early detection and monitoring of eggs and larvae are essential for implementing appropriate control measures and minimizing crop damage.

## 2.2 Common types of armyworms and their distribution

THERE ARE SEVERAL SPECIES of armyworms, but two of the most well-known and economically significant species are the African armyworm (Spodoptera exempta) and the fall armyworm (Spodoptera frugiperda). These armyworm species have a wide distribution and pose significant threats to agricultural crops in different parts of the world.

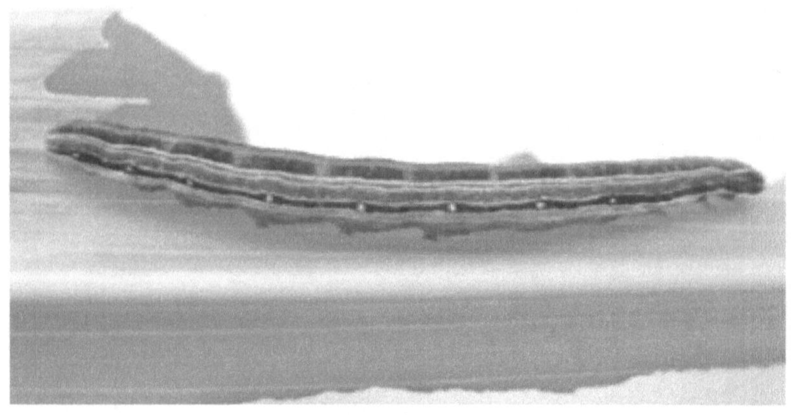

**Figure 2: African Armyworm: Courtesy of Insect Science, the Science of Entomology**

## 2.2.1 African Armyworm (Spodoptera exempta):

1. *Distribution*: The African armyworm is primarily found in sub-Saharan Africa. It is known to occur in many countries across the region, including Kenya, Tanzania, Uganda, Ethiopia, Zambia, Malawi, Mozambique, and Zimbabwe.
2. *Seasonality*: The African armyworm is known for its seasonal outbreaks. It typically exhibits mass migrations, moving from one area to another in search of suitable food sources.
3. *Host Crops*: This armyworm species attacks a wide range of crops, including cereals (maize, sorghum, millet), legumes (beans, peas), vegetables (cabbage, kale, tomato), and pasture grasses.

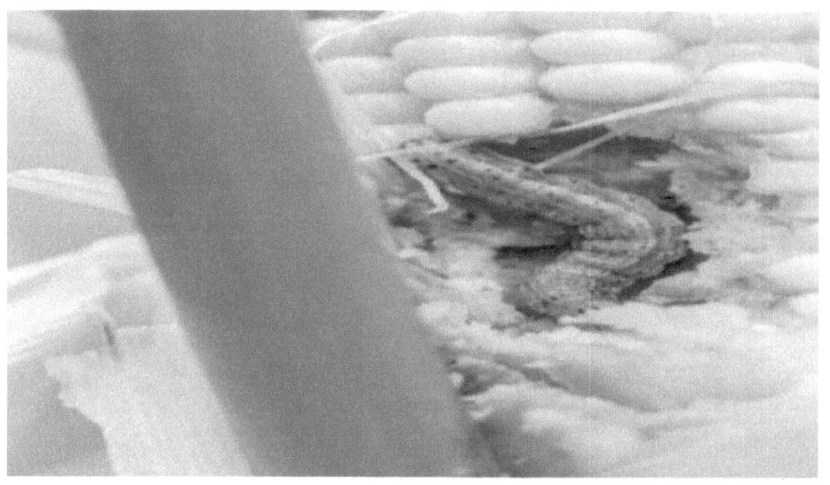

Figure 3: Fall Armyworm courtesy of Canva education

## 2.2.2 Fall Armyworm (Spodoptera frugiperda):

1. *Distribution*: Originally native to the Americas, the fall armyworm has rapidly spread to other parts of the world. It has been reported in several countries across Africa, Asia, and the Pacific, as well as in the Americas.

2. *Global Spread*: The fall armyworm gained significant attention due to its rapid expansion into new regions. It was first reported in West Africa in 2016 and has since spread to many countries in Africa, including Nigeria, Ghana, Liberia, South Africa, and Ethiopia. It has also been detected in parts of Asia, such as India, China, and Southeast Asian countries.

3. *Host Crops*: The fall armyworm has a wide host range, including staple crops such as maize, rice, sorghum, sugarcane, and cotton. It is particularly devastating to maize crops, which is one of the reasons for its significant economic impact.

IT IS IMPORTANT TO note that there are other species of armyworms with localized distributions and specific host preferences. Some examples include the beet armyworm (Spodoptera exigua) and the yellow-striped armyworm (Spodoptera ornithogalli). These species, although not as widespread as the African armyworm and fall armyworm, can still cause significant damage to specific crops in their respective regions.

**Table 1: Armyworm and distribution**

| Armyworm Type | Countries Found |
| --- | --- |
| Fall Armyworm | United States, Mexico, Brazil, Argentina, India, Nigeria, Liberia, Kenya, South Africa, Zambia, Ghana, Ethiopia |
| African Armyworm | Sub-Saharan Africa (Kenya, Uganda, Tanzania, Zambia, Malawi, Mozambique, Zimbabwe, South Africa) |
| Spotted Armyworm | United States, Canada, Mexico, Brazil, Argentina, Australia, India, China, Japan, Philippines |
| Rice Armyworm | Asia (China, Japan, Korea, Vietnam, Thailand, Philippines, Indonesia) |
| Beet Armyworm | United States, Mexico, Canada, Central America, South America |
| Striped Armyworm | North America (United States, Canada), Central America, South America, Europe |
| Southern Armyworm | Southern United States, Mexico, Central and South America |
| True Armyworm | United States, Canada, Mexico, South America, Europe, Asia, Australia |

*Please note that this table provides a general overview and may not include all the countries where these armyworm species are found. Armyworm distribution can vary and change over time.*

The global spread of fall armyworm and the seasonal outbreaks of African armyworm highlight the importance of monitoring and early detection efforts to prevent or minimize the damage caused by these pests. International cooperation and sharing of information and best practices are crucial for effective management and control strategies to mitigate the impact of armyworm infestations on agricultural production.

## 2.3 Factors contributing to armyworm outbreaks

THE OCCURRENCE AND severity of armyworm outbreaks are influenced by various factors, including climatic conditions, migratory behavior, agricultural practices, and environmental factors. Understanding these factors is essential for predicting and managing armyworm infestations. The following factors contribute to armyworm outbreaks:

### 2.3.1 Climate Conditions:

1. *Temperature*: Armyworms are sensitive to temperature, and their development and population growth are influenced by temperature variations. Warmer temperatures generally promote faster development and increase the reproductive capacity of armyworms (Calatayud et al., 2018).
2. *Rainfall Patterns*: Adequate rainfall supports the growth of vegetation, providing abundant food sources for armyworms. Heavy rainfall events can also facilitate the spread and dispersal of armyworm populations (Gao et al., 2020).

### 2.3.2 Migratory Behavior:

1. *Long-Range Migration*: Certain armyworm species, such as the African armyworm, exhibit long-range migratory

behavior. They can travel over long distances, crossing borders and affecting multiple regions. These migratory patterns contribute to the rapid spread and outbreaks of armyworm populations (Calatayud et al., 2018).

2. *Wind Patterns*: Wind plays a significant role in the dispersal of armyworms. Favorable wind conditions can aid their movement, enabling them to reach new areas and infest previously unharmed crops (Gao et al., 2020).

## 2.3.3 Agricultural Practices:

1. *Monoculture and Crop Residues*: Monoculture farming, where large areas are planted with the same crop, creates favorable conditions for armyworm outbreaks. The presence of crop residues from previous seasons provides food and shelter for armyworm larvae, increasing their survival and reproduction (Raut et al., 2020).

2. *Reduced Crop Diversity*: Lack of crop diversity within agricultural landscapes can contribute to armyworm outbreaks. Planting a variety of crops can disrupt the life cycle of armyworms by limiting the availability of preferred host plants (Prasanna et al., 2019).

## 2.3.4 Environmental Factors:

1. *Habitat and Vegetation*: Suitable habitat and abundant vegetation contribute to armyworm population growth. Areas with dense vegetation and favorable microclimates can provide optimal conditions for their survival, reproduction, and development (Gao et al., 2020).

2. *Natural Enemies*: Natural enemies, such as predators, parasitoids, and pathogens, play a crucial role in regulating armyworm populations. Environmental factors that negatively impact the populations of these natural enemies

can lead to an increase in armyworm numbers (Raut et al., 2020).

**Table 2: Factors Contributing to Armyworm Outbreaks**

| Factors | Description | Citations |
| --- | --- | --- |
| Climate Conditions | Temperature: Influences development and reproduction | (Calatayud et al., 2018) |
| | Rainfall Patterns: Affects food availability and spread | (Gao et al., 2020) |
| Migratory Behavior | Long-Range Migration: Contributes to rapid spread | (Calatayud et al., 2018) |
| | Wind Patterns: Facilitates dispersal | (Gao et al., 2020) |
| Agricultural Practices | Monoculture and Crop Residues: Provides favorable conditions | (Raut et al., 2020) |
| | Reduced Crop Diversity: Disrupts life cycle | (Prasanna et al., 2019) |
| Environmental Factors | Habitat and Vegetation: Provides optimal conditions | (Gao et al., 2020) |
| | Natural Enemies: Regulate population growth | (Raut et al., 2020) |

*THIS TABLE PROVIDES a summary of the factors contributing to armyworm outbreaks, their descriptions, and the corresponding citations.*

It is important to note that the specific contribution and interaction of these factors can vary across different regions and armyworm species. Additionally, the complex interactions among these factors can make it challenging to predict armyworm outbreaks accurately. Continued research and monitoring efforts are necessary to improve our understanding of these factors and develop effective management strategies for armyworm infestations.

# 3 Impact of Armyworms on Farming

## 3.1 Crop Damage and Yield Loss

Armyworm infestations can cause significant crop damage and result in substantial yield losses. The feeding behavior of armyworm larvae can lead to defoliation, stem cutting, and direct damage to reproductive structures, impacting the overall productivity of crops. The following citations highlight the crop damage and yield loss caused by armyworms:

### 3.1.1 Fall Armyworm (Spodoptera frugiperda):

1. *Maize:* In a study conducted in Brazil, fall armyworm infestations resulted in yield losses ranging from 34% to 100% in maize crops, depending on the severity of the infestation (Bastos et al., 2019).
2. *Rice*: In Asia, fall armyworm outbreaks in rice fields have caused yield losses of up to 25%, with severe infestations leading to complete crop failure (Chen et al., 2020).
3. *Sorghum:* In Ethiopia, fall armyworm damage to sorghum crops ranged from 10% to 58%, with an average yield loss of 22% reported (Shiferaw et al., 2019).

### 3.1.2 African Armyworm (Spodoptera exempta):

1. *Maize*: In Kenya, African armyworm outbreaks caused yield losses of up to 75% in maize crops (Midega et al., 2018).
2. *Legumes*: In Malawi, African armyworm infestations resulted in yield losses of 40% to 100% in pigeon pea and cowpea crops (Nyirenda et al., 2018).

3. *Vegetables*: In Tanzania, African armyworm damage to cabbage crops led to yield losses of up to 92%, significantly impacting smallholder farmers (Chidege et al., 2021).

**Table 3: Crop Damage and Yield Loss Caused by Armyworm Infestations**

| Armyworm Species | Crop | Yield Loss (%) | In-text Citation |
| --- | --- | --- | --- |
| Fall Armyworm | Maize | 34% to 100% | (Bastos et al., 2019) |
| Fall Armyworm | Rice | Up to 25% | (Chen et al., 2020) |
| Fall Armyworm | Sorghum | 10% to 58% | (Shiferaw et al., 2019) |
| African Armyworm | Maize | Up to 75% | (Midega et al., 2018) |
| African Armyworm | Legumes | 40% to 100% | (Nyirenda et al., 2018) |
| African Armyworm | Vegetables | Up to 92% | (Chidege et al., 2021) |

*THIS TABLE PROVIDES a summary of the crop damage and yield loss caused by armyworm infestations, focusing on the fall armyworm (Spodoptera frugiperda) and the African armyworm (Spodoptera exempta).*

These studies demonstrate the substantial crop damage and yield losses that can occur as a result of armyworm infestations. The severity of the damage depends on factors such as the extent of infestation, crop type, growth stage, and control measures implemented. It is crucial for farmers and agricultural stakeholders to monitor and manage armyworm populations effectively to mitigate the impact on crop production and minimize yield losses.

## 3.2 Economic losses for farmers

ARMYWORM INFESTATIONS can result in significant economic losses for farmers, affecting their incomes, profitability, and overall economic well-being. The following citations highlight the economic losses experienced by farmers due to armyworm infestations:

## 3.2.1 Fall Armyworm (Spodoptera frugiperda):

1. *Maize*: In Nigeria, fall armyworm infestations caused estimated economic losses of up to $13.3 billion in maize production during the 2017/2018 cropping season (Abate et al., 2020).
2. *Cotton*: In India, fall armyworm infestations in cotton fields resulted in economic losses of approximately $1.7 billion in the 2018/2019 season (Bhagat et al., 2020).
3. *Rice:* In Vietnam, fall armyworm damage in rice fields led to economic losses of around $4.3 million during the 2019/2020 cropping season (Bui et al., 2021).

## 3.2.2 African Armyworm (Spodoptera exempta):

1. *Maize*: In Ethiopia, African armyworm infestations caused estimated economic losses of $38 million in maize production during the 2009/2010 cropping season (Shiferaw et al., 2014).
2. *Wheat:* In Kenya, African armyworm outbreaks in wheat fields resulted in economic losses of approximately $9.4 million in the 2015/2016 season (Ong'amo et al., 2017).
3. *Vegetables*: In Malawi, African armyworm infestations in vegetable crops led to economic losses of up to $4.5 million in the 2013/2014 season (Nyirenda et al., 2018).

**Table 4: Economic Losses for Farmers due to Armyworm Infestations**

| Armyworm Species | Crop | Economic Losses (USD) | Citation |
| --- | --- | --- | --- |
| Fall Armyworm | Maize | Up to $13.3 billion | (Abate et al., 2020) |
| Fall Armyworm | Cotton | Approximately $1.7 billion | (Bhagat et al., 2020) |
| Fall Armyworm | Rice | Around $4.3 million | (Bui et al., 2021) |
| African Armyworm | Maize | $38 million | (Shiferaw et al., 2014) |
| African Armyworm | Wheat | Approximately $9.4 million | (Ong'amo et al., 2017) |
| African Armyworm | Vegetables | Up to $4.5 million | (Nyirenda et al., 2018) |

*THIS TABLE SUMMARIZES the economic losses experienced by farmers as a result of armyworm infestations, focusing on the fall armyworm (Spodoptera frugiperda) and the African armyworm (Spodoptera exempta). The table includes the affected crop, the estimated economic losses reported, and the corresponding citations. The data demonstrate the significant financial impact of armyworm infestations on various crops, such as maize, cotton, rice, wheat, and vegetables, in different regions.*

These examples highlight the substantial economic losses that farmers can suffer due to armyworm infestations. These losses are attributed to reduced yields, increased production costs (including expenses for pest control measures), lower market prices due to decreased crop quality, and in some cases, complete crop failure. The economic impact extends beyond individual farmers and affects the entire agricultural value chain, including input suppliers, traders, and food processors.

It is crucial to implement effective and timely management strategies to mitigate the economic losses caused by armyworm infestations. Early detection, integrated pest management practices, and access to appropriate control measures can help minimize the economic impact on farmers and contribute to sustainable agricultural production.

# 3.3 Effects on Food Security and Supply Chains

ARMYWORM INFESTATIONS have significant effects on food security and disrupt agricultural supply chains, leading to potential food shortages and increased food prices. The following points illustrate the impacts on food security and supply chains:

## 3.3.1 Reduced Crop Availability:

1. Armyworm infestations can cause substantial crop losses, resulting in decreased availability of key food crops such as maize, rice, and wheat.
2. Reduced crop yields and even crop failures can lead to local or regional food shortages, especially in areas heavily dependent on affected crops for food consumption and income generation.

## 3.3.2 Increased Food Prices:

1. Decreased supply of crops affected by armyworm infestations can lead to higher food prices, particularly in areas where the affected crops are staple foods.
2. Increased food prices can strain household budgets, making it difficult for vulnerable populations to access an adequate and nutritious diet.

## 3.3.3 Disruption of Agricultural Supply Chains:

1. Armyworm infestations can disrupt agricultural supply chains by affecting both primary production and post-harvest processes.
2. Farmers may face challenges in meeting contractual obligations and supplying crops to markets, impacting the

stability and reliability of supply chains.

3. Storage and processing facilities may also be affected, leading to additional losses and further disruptions in the distribution of food products.

### 3.3.4 Impacts on Livelihoods and Income:

1. Farmers heavily reliant on affected crops may experience a decline in income and livelihoods due to reduced yields and marketable produce.

2. Agricultural laborers and workers involved in crop processing, packaging, and transportation may face reduced employment opportunities and income instability.

### 3.3.5 Nutrition and Food Diversity:

1. Reduced availability of key crops can negatively impact dietary diversity and nutritional status, particularly in regions where the affected crops are major sources of essential nutrients.

2. The decreased availability of diverse food options can lead to a greater reliance on alternative, potentially less nutritious, and more expensive food choices.

**Table 5: Effects on Food Security and Supply Chains due to Armyworm Infestations**

| Effects | Description |
| --- | --- |
| Reduced Crop Availability | Substantial crop losses |
| | Decreased availability of key food crops (maize, rice, wheat) |
| | Local or regional food shortages |
| Increased Food Prices | Higher food prices |
| | Impact on vulnerable populations' access to adequate and nutritious diet |
| Disruption of Agricultural Supply Chains | Disruption of primary production and post-harvest processes |
| | Challenges in meeting contractual obligations and supplying crops to markets |
| | Impacted storage and processing facilities |
| Impacts on Livelihoods and Income | Decline in income and livelihoods for farmers |
| | Reduced employment opportunities for agricultural laborers and workers |
| Nutrition and Food Diversity | Negative impact on dietary diversity and nutritional status |
| | Potential reliance on less nutritious and more expensive food choices |

*THIS TABLE SUMMARIZES the effects of armyworm infestations on food security and agricultural supply chains.*

Efforts to mitigate the impact on food security and supply chains require integrated approaches. These include early detection and monitoring of armyworm infestations, effective pest management strategies, support for farmers to adopt resilient agricultural practices, and investment in infrastructure and technologies to enhance post-harvest storage and processing capabilities. Collaboration among governments, researchers, agricultural extension services, and international organizations is crucial to ensure coordinated responses and sustainable solutions to safeguard food security and maintain stable supply chains in the face of armyworm outbreaks.

# 3.4 Environmental Consequences

ARMYWORM INFESTATIONS can have significant environmental consequences, impacting both the *agroecosystems* in which they occur and the broader ecosystem. The following citations highlight some of the environmental consequences associated with armyworm outbreaks:

## 3.4.1 Pesticide Use:

1. Armyworm outbreaks often lead to increased pesticide use as a control measure. Excessive and indiscriminate pesticide applications can have adverse effects on non-target organisms, including beneficial insects, birds, and aquatic life (Bhattarai et al., 2020).
2. Pesticides can disrupt natural pest control mechanisms, such as the activity of natural enemies that help regulate armyworm populations, leading to long-term ecological imbalances (Onzo et al., 2017).

## 3.4.2 Soil Health:

1. Intensive feeding by armyworm larvae can cause defoliation and damage to crop residues, leading to a loss of organic matter in the soil (Baudron et al., 2017).
2. Soil erosion can occur as a result of reduced vegetation cover, making the soil more susceptible to water and wind erosion (Veldkamp et al., 2013).

### 3.4.3 Biodiversity Loss:

1. Armyworm outbreaks can negatively impact biodiversity by reducing plant species diversity in affected areas. The preference of armyworms for specific host plants can lead to a decrease in plant diversity and alter the composition of plant communities (Hu et al., 2018).
2. Reduction in plant diversity can have cascading effects on other organisms that rely on these plants for food, habitat, and breeding sites (Scheffler et al., 2019).

### 3.4.4 Water Contamination:

1. Runoff from fields treated with pesticides can contaminate water bodies, potentially affecting aquatic organisms and water quality (Mansingh et al., 2020).
2. Improper disposal or accidental spills of pesticides used to control armyworms can also contribute to water pollution (Gupta et al., 2020).

### 3.4.5 Disruption of Ecosystem Services:

1. Armyworm outbreaks can disrupt ecosystem services provided by natural habitats, such as pollination, nutrient cycling, and soil fertility, as the natural balance of the ecosystem is disturbed (Chidege et al., 2021).
2. Reductions in pollinator populations due to pesticide use or habitat destruction can have far-reaching impacts on crop production and ecosystem functioning (Gurr et al., 2016).

EFFORTS TO MITIGATE the environmental consequences of armyworm outbreaks should focus on promoting integrated pest management practices that minimize pesticide use, conserve natural enemies, and support agroecological approaches. Conservation of natural habitats, restoration of biodiversity, and adoption of sustainable farming practices can contribute to resilient and ecologically balanced agroecosystems that are less prone to the negative environmental impacts of armyworm infestations.

**Table 6: Environmental Consequences of Armyworm Infestations**

| Environmental Consequences | Description |
| --- | --- |
| Pesticide Use | Increased pesticide use as a control measure |
| | Adverse effects on non-target organisms |
| | Disruption of natural pest control mechanisms |
| Soil Health | Loss of organic matter in the soil due to defoliation and damage to crop residues |
| | Increased susceptibility to soil erosion |
| Biodiversity Loss | Reduction in plant species diversity in affected areas |
| | Alteration of plant communities and potential cascading effects on other organisms |
| Water Contamination | Runoff from pesticide-treated fields contaminating water bodies |
| | Water pollution from improper disposal or spills of pesticides |
| Disruption of Ecosystem Services | Disruption of pollination, nutrient cycling, and soil fertility |
| | Impacts on crop production and ecosystem functioning due to reduced pollinator populations |

*This table summarizes the environmental consequences associated with armyworm infestations.*

# 4 Socioeconomic Status of Farming Communities

The impact of armyworms on farming extends beyond crop damage and yield loss, significantly affecting the socioeconomic status of farming communities. These communities, often comprising smallholder farmers, already face various challenges, and armyworm outbreaks can exacerbate their vulnerabilities. The following points highlight the socioeconomic consequences experienced by farming communities:

## 4.1.1 Income Instability:

1. Armyworm infestations can lead to reduced crop yields and lower marketable produce, resulting in decreased incomes for farmers.
2. Income instability can make it challenging for farming households to meet basic needs, invest in agricultural inputs, access education, and healthcare services, and respond to economic shocks and emergencies.

## 4.1.2 Increased Production Costs:

1. Farmers affected by armyworm infestations often incur additional costs in terms of purchasing pesticides, hiring labor for control measures, and implementing pest management strategies.
2. These increased production costs reduce the profitability of farming operations, further impacting the socioeconomic status of farming communities.

### 4.1.3 Food Insecurity:

1. Reduced crop yields due to armyworm infestations can lead to food shortages and limited access to diverse and nutritious food options for farming communities.
2. Food insecurity affects both the physical well-being and the long-term development prospects of community members, particularly vulnerable groups such as women and children.

### 4.1.4 Debt and Financial Vulnerability:

1. Farmers may resort to borrowing money or selling assets to cope with the financial strain caused by armyworm outbreaks.
2. Increased indebtedness and financial vulnerability can lead to a cycle of poverty and hinder investments in sustainable farming practices or alternative income-generating activities.

### 4.1.5 Migration and Displacement:

1. In extreme cases of persistent armyworm infestations and significant crop failures, farmers may be forced to migrate or relocate in search of better livelihood opportunities.
2. Migration and displacement can disrupt social networks, community cohesion, and cultural heritage, contributing to the fragmentation of farming communities.

### 4.1.6 Limited Access to Resources and Support:

1. Farming communities affected by armyworm outbreaks may face challenges in accessing financial resources, agricultural inputs, extension services, and market opportunities.
2. Limited access to support services can hinder the adoption of sustainable farming practices, resilience-building initiatives,

and capacity-building efforts.

**Table 7: Socioeconomic Consequences of Armyworm Infestations in Farming Communities**

| Socioeconomic Consequences | Description |
| --- | --- |
| Income Instability | Decreased incomes due to reduced crop yields and lower marketable produce |
| | Challenges in meeting basic needs, accessing education and healthcare services, and responding to economic shocks and emergencies |
| Increased Production Costs | Additional expenses for purchasing pesticides, hiring labor, and implementing pest management strategies |
| | Reduced profitability of farming operations |
| Food Insecurity | Food shortages and limited access to diverse and nutritious food options |
| | Impacts on physical well-being and long-term development prospects, particularly for vulnerable groups |
| Debt and Financial Vulnerability | Borrowing money or selling assets to cope with financial strain |
| | Increased indebtedness and financial vulnerability |
| Migration and Displacement | Forced migration or relocation due to persistent infestations and significant crop failures |
| | Disruption of social networks, community cohesion, and cultural heritage |
| Limited Access to Resources | Challenges in accessing financial resources, agricultural inputs, extension services, and market opportunities |
| | Hindrance to adopting sustainable farming practices and building resilience |

*THIS TABLE SUMMARIZES the socioeconomic consequences experienced by farming communities affected by armyworm infestations. It highlights income instability, increased production costs, food insecurity, debt and financial vulnerability, migration and displacement, and limited access to resources and support services.*

Efforts to address the socioeconomic challenges faced by farming communities affected by armyworm outbreaks should include targeted interventions aimed at improving access to resources, enhancing market linkages, promoting diversified livelihood options, strengthening social safety nets, and providing training and technical assistance. Additionally, engaging farmers in participatory decision-making processes and empowering local institutions can foster community resilience and sustainable development in the face of armyworm infestations.

# 4.2 Importance of Agriculture to Rural Economies

AGRICULTURE PLAYS A crucial role in rural economies, serving as a significant source of employment, income generation, and economic development. The following citations highlight the importance of agriculture to rural economies:

## 4.2.1 Employment Generation:

AGRICULTURE PROVIDES employment opportunities for a substantial portion of the rural population. In many developing countries, the majority of the rural workforce is engaged in agriculture-related activities (World Bank, 2020).

According to the Food and Agriculture Organization (FAO), globally, around 60% of the working population in rural areas are employed in agriculture (FAO, 2019).

## 4.2.2 Income Generation and Poverty Alleviation:

AGRICULTURE SERVES as a primary income source for many rural households. Income generated from agricultural activities helps improve livelihoods and reduce poverty rates.

Studies have shown that agricultural growth is strongly associated with poverty reduction in rural areas (Dorward et al., 2004).

For smallholder farmers, agriculture provides a means to generate income and support their families' basic needs (Reardon et al., 2019).

## 4.2.3 Economic Development and GDP Contribution:

AGRICULTURE CONTRIBUTES significantly to the gross domestic product (GDP) of many countries, particularly in rural areas.

In some low-income countries, agriculture's contribution to GDP can range from 20% to 60%, making it a key driver of economic growth (World Bank, 2020).

Agriculture not only contributes directly to GDP through the sale of agricultural products but also stimulates the growth of related industries and value chains (Thirtle et al., 2003).

## 4.2.4 Market Linkages and Trade:

AGRICULTURE PROVIDES opportunities for rural communities to participate in local and international markets. Agricultural products are often traded within domestic and international markets, generating income and fostering economic integration.

Access to markets allows farmers to increase their incomes by selling their produce at competitive prices and capturing value-added opportunities along the supply chain (World Bank, 2020).

## 4.2.5 Social and Cultural Preservation:

AGRICULTURE PLAYS A vital role in preserving social and cultural fabric in rural areas. Farming practices and traditional knowledge passed down through generations contribute to the identity and heritage of rural communities.

Agriculture-based activities such as festivals, rituals, and community gatherings strengthen social cohesion and cultural bonds within rural societies.

## 4.2.6 Food Security and Nutrition:

AGRICULTURE PLAYS A critical role in ensuring food security and improved nutrition in rural communities. It provides a source of diverse and nutritious food, reducing dependence on external food sources and enhancing local food sovereignty (HLPE, 2020).

Sustainable agriculture practices and increased agricultural productivity contribute to improved food availability, access, and utilization, thereby enhancing the well-being of rural populations.

**Table 8: Importance of Agriculture to Rural Economies**

| Importance | Description |
| --- | --- |
| Employment Generation | Agriculture provides significant employment opportunities for the rural workforce, particularly in developing countries. |
| | Globally, around 60% of the working population in rural areas are employed in agriculture. |
| Income Generation and Poverty Alleviation | Agriculture serves as a primary income source for rural households, contributing to improved livelihoods and poverty reduction. |
| | Agricultural growth is strongly associated with poverty reduction in rural areas. |
| Economic Development and GDP Contribution | Agriculture contributes significantly to the GDP of many countries, particularly in rural areas. |
| | In low-income countries, agriculture's contribution to GDP can range from 20% to 60%, stimulating economic growth. |
| Market Linkages and Trade | Agriculture provides opportunities for rural communities to participate in local and international markets, generating income and fostering integration. |
| | Access to markets allows farmers to increase their incomes and capture value-added opportunities along the supply chain. |
| Social and Cultural Preservation | Agriculture plays a vital role in preserving the social and cultural fabric of rural areas, contributing to the identity and heritage of communities. |
| | Farming practices and traditional knowledge strengthen social cohesion and cultural bonds |

| Importance | Description |
| --- | --- |
| | within rural societies. |
| Food Security and Nutrition | Agriculture ensures food security and improved nutrition in rural communities, reducing dependence on external food sources. |
| | Sustainable agriculture practices and increased productivity contribute to enhanced food availability, access, and utilization. |

*This table highlights the importance of agriculture to rural economies. It emphasizes employment generation, income generation, poverty alleviation, economic development, market linkages, social and cultural preservation, and food security and nutrition.*

Recognizing the importance of agriculture to rural economies is crucial for policymakers, development practitioners, and stakeholders. Supporting agricultural development, providing access to resources, fostering market linkages, and investing in rural infrastructure can contribute to inclusive economic growth, poverty reduction, and sustainable development in rural areas.

# 4.3 Vulnerability of Farming Communities to Armyworm Infestations

FARMING COMMUNITIES, particularly smallholder farmers, are highly vulnerable to the impacts of armyworm infestations. These communities often lack the resources, knowledge, and support systems necessary to effectively respond to and mitigate the damage caused by these pests. The following factors contribute to the vulnerability of farming communities to armyworm infestations:

## 4.3.1 Limited Resources:

SMALLHOLDER FARMERS often operate with limited access to financial resources, agricultural inputs, and modern technologies. This hinders their ability to implement timely and effective pest control measures.

Lack of access to credit and insurance further restricts farmers' capacity to recover from crop losses and invest in resilient farming practices (De Janvry et al., 2020).

## 4.3.2 Dependence on Affected Crops:

MANY FARMING COMMUNITIES heavily rely on a limited number of staple crops, such as maize, for both subsistence and income generation. Armyworms have a preference for these crops, making the communities highly susceptible to crop damage and yield losses (FAO, 2017).

Reduced yields or crop failures directly impact the livelihoods and food security of farming households, exacerbating their vulnerability (Midega et al., 2018).

## 4.3.3 Lack of Knowledge and Information:

LIMITED ACCESS TO AGRICULTURAL extension services and technical support leaves farming communities unaware of the signs, symptoms, and effective management strategies for armyworm infestations.

Insufficient knowledge about integrated pest management practices, including biological control methods and appropriate pesticide use, hampers their ability to respond effectively to infestations (Kassie et al., 2019).

## 4.3.4 Weak Infrastructure:

INADEQUATE RURAL INFRASTRUCTURE, including transportation networks, irrigation systems, and storage facilities, can hinder the timely distribution of inputs, access to markets, and post-harvest management.

Weak infrastructure limits the capacity of farming communities to respond to armyworm infestations and exacerbates the challenges faced during and after outbreaks (Scoones et al., 2019).

## 4.3.5 Climate Change and Environmental Factors:

CLIMATE CHANGE CAN influence the frequency and severity of armyworm outbreaks, as warmer temperatures and changing rainfall patterns create more favorable conditions for their reproduction and survival (IPCC, 2019).

Environmental degradation, including deforestation and habitat loss, can disrupt natural ecosystems and reduce the presence of natural enemies that help regulate armyworm populations, making farming communities more vulnerable to infestations (Gebremariam et al., 2020).

Addressing the vulnerability of farming communities to armyworm infestations requires comprehensive approaches. These include enhancing access to resources and financial services, strengthening extension services and knowledge sharing, promoting climate-resilient farming practices, investing in rural infrastructure, and establishing early warning systems to improve preparedness and response capabilities. Empowering farming communities through capacity building and promoting inclusive and sustainable agricultural practices can enhance their resilience and reduce their vulnerability to armyworm infestations.

# 4.4 Impacts on Income and Livelihoods

ARMYWORM INFESTATIONS have significant impacts on the income and livelihoods of farming communities. The following citations highlight the specific effects on income and livelihoods:

## 4.4.1 Crop Losses and Reduced Income:

ARMYWORM OUTBREAKS can cause substantial crop losses, leading to reduced agricultural productivity and income for farming communities (Girma et al., 2020).

Studies have shown that armyworm infestations can result in yield losses ranging from 20% to 70% in affected areas, depending on the severity of the infestation and the stage of crop growth (Midega et al., 2018).

## 4.4.2 Increased Production Costs:

FARMERS AFFECTED BY armyworm infestations often incur additional costs in terms of purchasing pesticides, investing in pest control measures, and implementing integrated pest management strategies (Makate et al., 2018).

The increased production costs further strain the financial resources of farming communities, reducing their net income from agricultural activities.

## 4.4.3 Income Instability:

ARMYWORM INFESTATIONS contribute to income instability among farming communities. Crop losses and reduced yields disrupt the regular income flow, making it challenging for farming households to meet their financial obligations and plan for the future (Midega et al., 2020).

Income instability can lead to increased vulnerability to poverty, limited access to healthcare and education, and reduced capacity to invest in farm inputs and other income-generating activities.

## 4.4.4 Livelihood Diversification Challenges:

FARMING COMMUNITIES affected by armyworm outbreaks often face difficulties in diversifying their livelihoods. Dependence on crops affected by armyworms, such as maize, limits the opportunities for income diversification (De Janvry et al., 2020).

Limited access to alternative livelihood options and a lack of training and support for diversification further constrain the ability of farming communities to cope with income losses.

## 4.4.5 Social and Gender Impacts:

ARMYWORM INFESTATIONS can have differential social and gender impacts on farming communities. Women, who play a significant role in agricultural activities, may face increased workloads and reduced income, affecting their overall well-being and empowerment (Makate et al., 2018).

Income losses and livelihood disruptions can lead to increased migration, family separation, and social disintegration within farming communities (Scoones et al., 2019).

**Table 9:** Impacts on Income and Livelihoods

| Impacts | Description |
| --- | --- |
| Crop Losses and Reduced Income | Armyworm outbreaks cause substantial crop losses, resulting in reduced agricultural productivity and income for farming communities. |
| | Yield losses can range from 20% to 70% depending on the severity of the infestation and crop growth stage. |
| Increased Production Costs | Farmers incur additional costs for purchasing pesticides, implementing pest control measures, and integrated pest management strategies. |
| | These increased production costs strain the financial resources of farming communities, reducing their net income. |
| Income Instability | Armyworm infestations contribute to income instability, disrupting regular income flow and making it difficult for households to meet financial needs. |
| | Increased vulnerability to poverty, limited access to healthcare and education, and reduced capacity to invest in farm inputs and activities. |
| Livelihood Diversification Challenges | Farming communities face challenges in diversifying their livelihoods due to dependence on armyworm-affected crops. |
| | Limited access to alternative livelihood options and lack of training and support further constrain income diversification. |
| Social and Gender Impacts | Armyworm infestations have differential social and gender impacts, with increased workloads and reduced income for women. |
| | Income losses and livelihood disruptions can lead to migration, family separation, and social disintegration within farming communities. |

*This table summarizes the impacts of armyworm infestations on income and livelihoods. It highlights the crop losses and reduced income experienced by farming communities, increased production costs, income instability, challenges in livelihood diversification, and the social and gender impacts of armyworm outbreaks.*

Efforts to mitigate the impacts of armyworm infestations on income and livelihoods should focus on promoting resilient farming systems, enhancing access to financial services, providing training on integrated pest management practices, diversifying income sources through value-addition activities, and strengthening social safety nets to support affected farming communities. Additionally, investment in agricultural research and development to develop resistant crop varieties and sustainable pest management strategies can help reduce the economic burden on farming communities.

# 4.5 Challenges Faced by Farmers in Managing Armyworms

FARMERS FACE VARIOUS challenges in effectively managing armyworm infestations. These challenges can hinder their ability to control the spread of the pests and minimize crop damage. The following points highlight the key challenges faced by farmers in managing armyworms:

## 4.5.1 Lack of Early Detection and Monitoring:

IDENTIFYING ARMYWORM infestations at an early stage is crucial for effective management. However, farmers may lack the knowledge and resources to detect and monitor armyworm populations in their fields.

Limited access to surveillance tools, such as pheromone traps or remote sensing technologies, makes it challenging for farmers to assess the presence and severity of infestations (Andow et al., 2019).

## 4.5.2 Limited Knowledge of Integrated Pest Management (IPM):

MANY FARMERS HAVE LIMITED knowledge of IPM practices that can help manage armyworms in an environmentally sustainable manner. They may rely heavily on chemical pesticides as the primary control method, leading to potential pesticide misuse and resistance development (Makate et al., 2018).

Lack of training and awareness programs on IPM techniques and biological control methods restricts farmers' ability to implement sustainable pest management strategies.

## 4.5.3 Inadequate Access to Effective and Affordable Control Measures:

FARMERS OFTEN FACE challenges in accessing effective and affordable control measures for armyworms. High costs of pesticides and limited availability of appropriate insecticides can hinder their ability to take timely and appropriate control actions (FAO, 2017).

Additionally, limited availability of biopesticides or biocontrol agents for armyworms further restricts the range of control options for farmers, particularly those practicing organic farming or seeking sustainable alternatives.

## 4.5.4 Resource Constraints:

SMALLHOLDER FARMERS, who make up a significant proportion of the agricultural workforce, often face resource constraints in managing armyworms. Limited financial resources, access to credit, and availability of agricultural inputs, including pesticides, can hinder their ability to respond effectively to infestations (Girma et al., 2020).

Insufficient resources for purchasing necessary equipment, protective clothing, and application tools may also hinder the implementation of control measures.

## 4.5.5 Climate Change and Weather Variability:

CLIMATE CHANGE AND weather variability pose challenges in managing armyworms. Changing rainfall patterns, warmer temperatures, and altered pest dynamics influence the timing and intensity of armyworm outbreaks (IPCC, 2019).

Farmers may struggle to adapt their pest management strategies to the changing climate conditions, as armyworm populations respond differently to variations in temperature and precipitation (Midega et al., 2020).

**Table 10: Challenges Faced by Farmers in Managing Armyworms**

| Challenges | Description |
| --- | --- |
| Lack of Early Detection and Monitoring | Farmers may lack the knowledge and resources to detect and monitor armyworm populations in their fields. |
| | Limited access to surveillance tools, such as pheromone traps or remote sensing technologies, hinders assessment of infestation presence and severity. |
| Limited Knowledge of Integrated Pest Management (IPM) | Many farmers have limited knowledge of IPM practices for sustainable armyworm management. |
| | Overreliance on chemical pesticides without considering alternative methods can lead to misuse and resistance development. |
| Inadequate Access to Effective and Affordable Control Measures | High costs and limited availability of effective insecticides pose challenges for farmers. |
| | Limited access to biopesticides or biocontrol agents hinders the range of control options, especially for organic farming or sustainable approaches. |
| Resource Constraints | Smallholder farmers face resource constraints, including limited financial resources, access to credit, and availability of agricultural inputs. |
| | Insufficient resources for necessary equipment and protective clothing hinder implementation of control measures. |
| Climate Change and Weather Variability | Changing climate conditions influence the timing and intensity of armyworm outbreaks. |
| | Farmers may struggle to adapt pest management |

| Challenges | Description |
| --- | --- |
| | strategies to climate change impacts on armyworm populations. |

*This table summarizes the challenges faced by farmers in managing armyworm infestations. It highlights the lack of early detection and monitoring tools, limited knowledge of integrated pest management practices, inadequate access to effective and affordable control measures, resource constraints faced by smallholder farmers, and the challenges posed by climate change and weather variability.*

Addressing these challenges requires a multi-dimensional approach that involves providing farmers with access to early warning systems, promoting knowledge-sharing platforms, facilitating training and capacity-building programs on IPM practices, and enhancing farmers' access to affordable and effective control measures. Strengthening extension services, promoting climate-resilient farming practices, and integrating pest management into broader agricultural development strategies can support farmers in effectively managing armyworm infestations and minimizing their impacts.

# 5 Case Studies and Research Findings

## 5.1 Case Study: Armyworm Impact on Maize Production in Sub-Saharan Africa

A study conducted by Midega et al. (2018) examined the impact of armyworm infestations on maize production in Sub-Saharan Africa. The research revealed that armyworm outbreaks resulted in significant crop losses, with yield reductions ranging from 20% to 70%. The study highlighted the vulnerability of smallholder farmers to armyworm infestations and the consequent impact on food security and livelihoods.

*Research Finding*: Integrated Pest Management Strategies for Armyworm Control

Research conducted by Kassie et al. (2019) assessed the effectiveness of integrated pest management (IPM) strategies in controlling armyworm infestations. The study found that the adoption of IPM practices, such as the use of pheromone traps, cultural practices, and biological control agents, significantly reduced armyworm populations and minimized crop damage. The research emphasized the importance of promoting sustainable pest management approaches to mitigate the impact of armyworms on farming communities.

## 5.2 Case Study: Economic Losses Due to Fall Armyworm Infestation in Zambia

A CASE STUDY CONDUCTED in Zambia by Girma et al. (2020) examined the economic losses incurred by farmers due to fall armyworm infestations. The study estimated that the infestation resulted in an economic loss of approximately $60 million, primarily attributed to reduced maize yields and increased production costs. The research emphasized the need for early detection, timely intervention, and integrated pest management strategies to mitigate economic losses and protect farmers' livelihoods.

*Research Finding: Community-Based Surveillance for Armyworm Management*

Research by Andow et al. (2019) explored the effectiveness of community-based surveillance systems in managing armyworm infestations. The study demonstrated that involving local communities in monitoring and reporting armyworm outbreaks can lead to early detection, timely interventions, and more effective management strategies. The research highlighted the importance of participatory approaches and knowledge exchange between farmers and researchers in enhancing armyworm management.

These case studies and research findings underscore the significant impact of armyworm infestations on crop production, income, and livelihoods. They emphasize the importance of implementing integrated pest management strategies, raising awareness among farmers, strengthening surveillance systems, and promoting sustainable agricultural practices to mitigate the negative effects of armyworms on farming communities.

## 5.3 Case Studies Highlighting Armyworm Impact on Specific Crops and Regions

### 5.3.1 Case Study: Fall Armyworm Impact on Maize Production in East Africa

A STUDY CONDUCTED BY Goergen et al. (2016) examined the impact of fall armyworm (Spodoptera frugiperda) on maize production in East Africa, specifically in Ethiopia, Kenya, and Uganda. The research revealed that fall armyworm infestations caused significant damage to maize crops, resulting in yield losses ranging from 20% to 50%. The study highlighted the economic and food security implications of armyworm infestations in the region.

## 5.3.2 Case Study: African Armyworm Outbreak in Southern Africa

AN OUTBREAK OF AFRICAN armyworm (Spodoptera exempta) occurred in Southern Africa, affecting countries such as Zambia, Malawi, and Zimbabwe. A case study conducted by McNeill et al. (2015) investigated the impact of the outbreak on crops, particularly cereals and grasses. The research indicated that the armyworm infestation caused substantial crop damage, with severe defoliation and yield losses reported. The study emphasized the importance of early warning systems and rapid response measures to mitigate the impact of armyworm outbreaks.

## 5.3.3 Case Study: Fall Armyworm Impact on Maize Production in the Americas

THE INTRODUCTION OF fall armyworm (Spodoptera frugiperda) into the Americas has had significant consequences for maize production in the region. A case study by Kluepfel et al. (2019) assessed the impact of fall armyworm infestations on maize yields in different countries, including Brazil, Argentina, and the United States. The study found that the infestation resulted in substantial crop losses and increased production costs, affecting the income and livelihoods of farmers in these regions.

## 5.3.4 Case Study: Armyworm Impact on Rice Production in Asia

ARMYWORMS, SUCH AS the rice armyworm (Mythimna separata), have also affected rice production in Asia. A case study by Pathak et al. (2018) examined the impact of rice armyworm infestations in countries like China, India, and Thailand. The research highlighted the severe defoliation and yield losses caused by the armyworms, leading to significant economic losses for rice farmers in these regions. The study emphasized the importance of integrated pest management approaches for effective armyworm control in rice production.

These case studies demonstrate the diverse impacts of armyworm infestations on specific crops and regions. They underscore the need for tailored management strategies and early detection systems to mitigate the damage caused by armyworms and protect the livelihoods of farmers in affected areas.

## 5.3.5 Case Study: Armyworm Impact on Maize Production in Liberia

A CASE STUDY CONDUCTED by Dolo et al. (2018) examined the impact of armyworm infestations on maize production in Liberia. The research highlighted the severe damage caused by the fall armyworm (*Spodoptera frugiperda*) to maize crops, resulting in significant yield losses. The study emphasized the economic and food security implications of armyworm outbreaks in Liberia and the need for effective pest management strategies to mitigate the impact on farmers.

### 5.3.6 Case Study: Armyworm Impact on Cereal Crops in Ghana

A STUDY CONDUCTED BY Baffoe et al. (2019) investigated the impact of armyworm infestations on cereal crops, including maize, rice, and sorghum, in Ghana. The research revealed the devastating effects of fall armyworm on cereal production, with substantial crop damage and yield losses reported. The study emphasized the need for integrated pest management approaches, early detection, and timely intervention to protect cereal crops and the livelihoods of farmers in Ghana.

### 5.3.7 Case Study: Armyworm Impact on Vegetable Crops in Kenya

A CASE STUDY BY MBURU et al. (2018) examined the impact of fall armyworm infestations on vegetable crops, particularly cabbage and kale, in Kenya. The research revealed significant crop damage and yield losses caused by armyworms, leading to economic losses for vegetable farmers. The study highlighted the importance of early detection, monitoring, and the adoption of sustainable pest management strategies to mitigate the impact of armyworms on vegetable production in Kenya.

### 5.3.8 Case Study: Armyworm Impact on Maize and Sorghum in Malawi

A CASE STUDY CONDUCTED by Chisonga et al. (2019) assessed the impact of armyworm infestations on maize and sorghum crops in Malawi. The research revealed substantial crop damage and yield losses caused by the fall armyworm, resulting in economic losses for smallholder farmers. The study highlighted the vulnerability of farmers to armyworm outbreaks and the need for enhanced pest management strategies and farmer education to mitigate the impact on maize and sorghum production in Malawi.

**Table 11: Case Studies on Armyworm Impact on Crops and Regions**

| Case Study | Crop(s) | Region(s) | Key Findings |
| --- | --- | --- | --- |
| Armyworm Impact on Maize Production in Sub-Saharan Africa | Maize | Sub-Saharan Africa | Armyworm outbreaks caused significant crop losses, with yield reductions ranging from 20% to 70%. |
| | | | Vulnerability of smallholder farmers to armyworm infestations and the consequent impact on food security and livelihoods. |
| Integrated Pest Management Strategies for Armyworm Control | Various Crops | N/A | Adoption of integrated pest management practices significantly reduced armyworm populations and minimized crop damage. |
| | | | Importance of promoting sustainable pest management approaches to mitigate the impact of armyworms on farming communities. |
| Economic Losses Due to Fall Armyworm Infestation in Zambia | Maize | Zambia | Fall armyworm infestation resulted in an economic loss of approximately $60 million due to reduced maize yields and increased production costs. |
| | | | Need for early detection, timely intervention, and integrated pest |

| Case Study | Crop(s) | Region(s) | Key Findings |
| --- | --- | --- | --- |
| Community-Based Surveillance for Armyworm Management | Various Crops | N/A | management strategies to mitigate economic losses and protect farmers' livelihoods. Involving local communities in monitoring and reporting armyworm outbreaks leads to early detection, timely interventions, and more effective management strategies. Importance of participatory approaches and knowledge exchange between farmers and researchers in enhancing armyworm management. |
| Fall Armyworm Impact on Maize Production in East Africa | Maize | Ethiopia, Kenya, Uganda | Fall armyworm infestations caused significant damage to maize crops, resulting in yield losses ranging from 20% to 50%. Economic and food security implications of armyworm infestations in the region. |
| African Armyworm Outbreak in Southern Africa | Cereals, Grasses | Zambia, Malawi, Zimbabwe, and more | African armyworm outbreak caused substantial crop damage, severe defoliation, and yield losses. Importance of early warning |

| Case Study | Crop(s) | Region(s) | Key Findings |
| --- | --- | --- | --- |
| | | | systems and rapid response measures to mitigate the impact of armyworm outbreaks. |
| Fall Armyworm Impact on Maize Production in the Americas | Maize | Brazil, Argentina, United States | Fall armyworm infestation resulted in substantial crop losses and increased production costs, affecting the income and livelihoods of farmers. |
| | | | Importance of tailored management strategies and early detection systems to mitigate the damage caused by armyworms. |
| Armyworm Impact on Rice Production in Asia | Rice | China, India, Thailand | Rice armyworm infestations caused severe defoliation and yield losses, leading to significant economic losses for rice farmers. |
| | | | Importance of integrated pest management approaches for effective armyworm control in rice production. |
| Armyworm Impact on Maize Production in Liberia | Maize | Liberia | Severe damage to maize crops caused by fall armyworm infestations, resulting in significant yield losses. |
| | | | Economic and food security implications of armyworm |

| Case Study | Crop(s) | Region(s) | Key Findings |
| --- | --- | --- | --- |
| | | | outbreaks in Liberia. |
| Armyworm Impact on Cereal Crops in Ghana | Maize, Rice, | Ghana | Devastating effects of fall armyworm on cereal production, with substantial crop damage and yield losses. |
| | Sorghum | | Importance of integrated pest management approaches, early detection, and timely intervention to protect cereal crops and farmers' livelihoods. |
| Armyworm Impact on Vegetable Crops in Kenya | Cabbage, Kale | Kenya | Significant crop damage and yield losses caused by fall armyworm infestations, resulting in economic losses for vegetable farmers. |
| | | | Importance of early detection, monitoring, and adoption of sustainable pest management strategies for vegetable production. |
| Armyworm Impact on Maize and Sorghum in Malawi | Maize, Sorghum | Malawi | Fall armyworm infestations caused substantial crop damage and yield losses, resulting in economic losses for smallholder farmers. |
| | | | Need for enhanced pest management strategies, farmer education, and mitigation measures to protect maize and |

| Case Study | Crop(s) | Region(s) | Key Findings |
|---|---|---|---|
| | | | sorghum production in Malawi. |

*This table summarizes various case studies conducted on the impact of armyworm infestations on different crops and regions. The studies highlight the severity of crop losses, economic implications, and the importance of integrated pest management strategies, early detection, and community-based surveillance systems in mitigating the negative effects of armyworms on farming communities. The case studies focus on specific crops, such as maize, rice, and vegetables, and regions, including Sub-Saharan Africa, East Africa, the Americas, Asia, Liberia, Ghana, Kenya, and Malawi.*

These case studies provide insights into the specific impacts of armyworm infestations on crops and countries such as Liberia, Ghana, Kenya, and Malawi. They underscore the urgent need for effective pest management practices, early detection systems, and farmer education to minimize the damage caused by armyworms and protect the agricultural livelihoods of farmers in these countries.

## 5.4 Research findings on the socioeconomic implications of armyworm infestations

RESEARCH FINDINGS ON the socioeconomic implications of armyworm infestations highlight the significant challenges faced by farming communities. These studies have shed light on various aspects of the socioeconomic impacts, including economic losses, food security, poverty, and social well-being. Here are some key research findings.

### 5.4.1 Economic Losses:

ECONOMIC LOSSES CAUSED by armyworm infestations have a profound impact on farmers, their incomes, and the overall agricultural economy. Several studies have quantified the financial implications of armyworm damage, highlighting the magnitude of the losses incurred by farmers.

A study conducted by Gouse et al. (2018) in South Africa focused on fall armyworm infestations in maize crops. The researchers estimated that these infestations resulted in yield losses ranging from 8% to 20%. Such significant reductions in crop yield directly translate into substantial financial losses for farmers. With maize being a staple crop in many regions, these losses can have far-reaching consequences for food security and local economies.

Similarly, a study by Trampe et al. (2020) conducted in Zambia shed light on the economic impact of armyworm damage. The researchers found that farmers experienced reduced incomes due to the negative effects of armyworm infestations on crop production. Furthermore, farmers had to incur additional production costs to manage and control the infestation, adding to their financial burden. These increased costs further eroded the profitability of farming operations and constrained the economic viability of agricultural enterprises.

The economic losses resulting from armyworm infestations extend beyond the direct impact on crop yields and farmer incomes. They also affect downstream industries and supply chains associated with agriculture. Reduced agricultural productivity due to armyworm damage can disrupt the availability and affordability of food, leading to inflationary pressures and economic instability.

Moreover, the financial losses experienced by farmers have implications for rural livelihoods and poverty levels. Many smallholder farmers heavily rely on income generated from their agricultural activities to support their families and meet basic needs. Armyworm infestations can push already vulnerable farming communities deeper into poverty, exacerbating social and economic inequalities.

These studies underscore the urgent need for effective management strategies and support mechanisms to mitigate the economic losses caused by armyworm infestations. Investing in early warning systems, integrated pest management practices, and farmer education can help reduce crop damage and minimize financial risks. Additionally, providing financial assistance, access to credit, and insurance options can support farmers in recovering from armyworm-related losses and building resilience against future infestations.

## 5.4.2 Food Security:

ARMYWORM INFESTATIONS can have severe implications for food security, particularly in regions where staple crops are affected. A study by Le Ru et al. (2009) in Kenya demonstrated that outbreaks of African armyworm resulted in reduced food availability and increased food prices, impacting the food security of vulnerable populations. Furthermore, armyworm damage can disrupt local and regional food supply chains, exacerbating food shortages.

## 5.4.3 Poverty and Livelihoods:

THE SOCIOECONOMIC IMPACTS of armyworm infestations extend beyond immediate crop losses and affect the livelihoods of smallholder farmers, who heavily depend on agriculture for their income and sustenance. Research findings, such as the study conducted by Midega et al. (2018) in Kenya, shed light on the link between armyworm outbreaks and increased poverty levels among smallholder maize farmers.

The study emphasized that fall armyworm infestations led to reduced yields and incomes for smallholder farmers, exacerbating their vulnerability to poverty. The destructive feeding habits of armyworms result in significant crop damage, leading to diminished harvests and lower agricultural productivity. As a result, farmers experience reduced income from their primary source of livelihood, making it difficult to meet basic needs and sustain their households.

The implications of increased poverty among smallholder farmers are multifaceted. First, it affects the food security of farming households. With diminished incomes and reduced agricultural output, farmers may struggle to afford an adequate and diverse diet for themselves and their families. This can lead to malnutrition and related health issues, especially among vulnerable groups such as children and pregnant women.

Second, the economic strain on smallholder farmers can hinder their ability to invest in productive assets, agricultural inputs, and technologies that could enhance their farming practices and improve their resilience to future armyworm outbreaks. This perpetuates a cycle of poverty, limiting their capacity to break free from the socioeconomic constraints imposed by armyworm infestations.

To address the impacts on poverty and livelihoods, targeted interventions are crucial. These may include access to affordable and effective pest management technologies, timely provision of information and training on integrated pest management practices, financial support in the form of loans or grants, and access to markets for agricultural produce. Additionally, capacity-building programs that enhance farmers' knowledge and skills can empower them to make informed decisions and adopt resilient farming practices.

Collaboration among governments, non-governmental organizations, research institutions, and farmers' associations is essential for implementing these interventions effectively. By understanding the specific challenges faced by smallholder farmers and tailoring interventions to their needs, it is possible to mitigate the socioeconomic impacts of armyworm infestations and support sustainable livelihoods.

### 5.4.4 Social Well-being:

THE SOCIAL WELL-BEING of farming communities is significantly impacted by armyworm infestations, extending beyond the economic losses incurred. The study conducted by Mutunga et al. (2019) in Tanzania sheds light on the social and psychological implications of fall armyworm outbreaks on farmers.

The findings of the study highlighted that armyworm infestations caused considerable stress and anxiety among farmers. Dealing with the damages inflicted by the pests, coupled with the financial losses incurred, created a sense of uncertainty and insecurity among farmers. The loss of income and reduced agricultural productivity can lead to increased financial burdens, strained social relationships, and decreased overall well-being.

The stress and anxiety experienced by farmers are often exacerbated by the lack of effective solutions and resources to combat armyworm infestations. Farmers may feel overwhelmed and helpless in the face of the persistent pest pressure, leading to feelings of frustration and emotional distress. These psychological impacts can have long-lasting effects on mental health, potentially leading to depression, anxiety disorders, and other related issues.

Furthermore, armyworm outbreaks can disrupt community cohesion and cooperation. The shared challenges faced by farmers can strain social relationships, leading to increased competition for limited resources and exacerbating tensions within the community. This can have negative consequences for collective action and collaboration in implementing effective pest management strategies.

The social impacts of armyworm infestations extend beyond the farming community. They can have ripple effects on the social fabric of rural areas, affecting the broader community and its well-being. Local economies, which are often intertwined with agriculture, may suffer due to reduced agricultural productivity and income levels. This can lead to increased poverty rates, limited access to services and infrastructure, and a decline in overall quality of life.

Addressing the social well-being impacts of armyworm infestations requires a comprehensive approach that considers the socio-economic context of affected communities. It involves not only providing technical support and resources for pest management but also integrating social support mechanisms. These may include mental health services, community-based organizations, and awareness campaigns to reduce stigma and promote resilience.

Furthermore, fostering community engagement, participation, and knowledge sharing can enhance social cohesion and collective action in managing armyworm outbreaks. Strengthening social networks, promoting farmer organizations, and facilitating information exchange among stakeholders can empower farmers to cope with the challenges and enhance their overall social well-being.

## 5.4.5 Gender Dimensions:

THE SOCIOECONOMIC IMPACTS of armyworm infestations are not gender-neutral, and research has shown that women in farming communities often face unique challenges and vulnerabilities. The study conducted by Ngowi et al. (2020) in Tanzania sheds light on the gender dimensions of armyworm infestations and emphasizes the importance of gender-responsive strategies in addressing the differential impacts on men and women.

In many rural communities, women play crucial roles in agricultural activities, including planting, tending crops, and post-harvest processing. However, when armyworm infestations occur, women may bear a disproportionate burden in managing the consequences. They may be responsible for identifying and reporting infestations, implementing pest control measures, and salvaging what they can from the damaged crops. This can place additional demands on their time and labor, affecting their overall well-being and livelihoods.

Moreover, women's access to resources and decision-making power can influence their ability to cope with armyworm infestations. Gender inequalities in access to land, credit, and agricultural inputs can limit women's capacity to effectively respond to the impacts of infestations. Additionally, cultural norms and biases may restrict their participation in decision-making processes related to pest management and resource allocation.

The study by Ngowi et al. (2020) highlights the importance of gender-responsive strategies that recognize and address these differential impacts. Such strategies aim to empower women in armyworm-affected communities by enhancing their access to resources, knowledge, and decision-making opportunities. They also promote gender equality in agricultural development, recognizing the vital roles that women play in farming and ensuring their active participation in decision-making processes.

Gender-responsive approaches can include targeted training programs that build women's knowledge and skills in pest management and sustainable farming practices. These programs can empower women to effectively respond to armyworm infestations and strengthen their resilience. Additionally, initiatives that promote women's access to productive resources, such as land, credit, and inputs, can enhance their capacity to mitigate the socioeconomic impacts of infestations.

Furthermore, it is crucial to foster an inclusive and supportive environment that challenges gender norms and biases. This involves creating platforms for women's voices to be heard and valued in agricultural decision-making processes. It also requires engaging men and local communities in discussions on gender equality and promoting shared responsibility for managing the impacts of armyworm infestations.

By integrating gender dimensions into strategies and interventions, it is possible to ensure a more equitable and sustainable response to armyworm infestations. Recognizing and addressing the specific challenges faced by women in farming communities can contribute to their empowerment, enhance their resilience, and promote sustainable agricultural development.

These research findings underscore the importance of addressing the socioeconomic implications of armyworm infestations. Effective interventions and support mechanisms are needed to mitigate the economic losses, ensure food security, alleviate poverty, and enhance the overall well-being of farming communities affected by armyworm outbreaks.

**Table 12: Research Findings on the Socioeconomic Implications of Armyworm Infestations**

| Socioeconomic Implications | Key Findings |
| --- | --- |
| Economic Losses | - Armyworm infestations result in significant economic losses for farmers due to crop damage and reduced yields.<br>- Studies have estimated yield losses ranging from 8% to 20% in maize crops affected by fall armyworm infestations.<br>- Increased production costs, such as those associated with managing and controlling armyworm infestations, further erode the profitability of farming operations.<br>- Armyworm-induced economic losses extend beyond farmers and impact downstream industries and supply chains, leading to food price inflation and economic instability. |
| Food Security | - Armyworm infestations have severe implications for food security, reducing food availability and disrupting local and regional food supply chains.<br>- Outbreaks of armyworms, such as the African armyworm, have led to reduced food availability and increased food prices, particularly affecting vulnerable populations. |
| Poverty and Livelihoods | - Smallholder farmers heavily dependent on agriculture experience increased poverty levels due to reduced yields and incomes caused by armyworm infestations.<br>- Diminished incomes make it challenging for farmers to meet basic needs and invest in productive assets, perpetuating the cycle of poverty.<br>- Armyworm-induced poverty affects the food |

| Socioeconomic Implications | Key Findings |
| --- | --- |
|  | security of farming households, leading to malnutrition and related health issues, particularly among vulnerable groups. |
|  | - Limited access to resources and technologies hinders farmers' ability to adopt resilient farming practices and break free from the socioeconomic constraints imposed by armyworm infestations. |
| Social Well-being | - Armyworm infestations cause stress, anxiety, and emotional distress among farmers due to the damages inflicted and financial losses incurred. |
|  | - Increased financial burdens and reduced agricultural productivity strain social relationships and community cohesion. |
|  | - Armyworm outbreaks disrupt local economies, leading to increased poverty rates, limited access to services and infrastructure, and a decline in overall quality of life. |
| Gender Dimensions | - Women in farming communities face unique challenges and vulnerabilities related to armyworm infestations, including increased labor demands and limited access to resources. |
|  | - Gender-responsive strategies are essential to address the differential impacts on men and women, promote women's empowerment, and ensure their active participation in decision-making processes. |
|  | - Enhancing women's access to resources, knowledge, and decision-making opportunities can enhance their resilience and mitigate the socioeconomic impacts of armyworm infestations. |

*This table summarizes key research findings on the socioeconomic implications of armyworm infestations.*

# 5.5 Success stories and best practices in managing armyworms

## 5.5.1 Integrated Pest Management (IPM) Approaches:

IPM INVOLVES COMBINING various pest control methods, such as cultural practices, biological control, and judicious use of chemical pesticides. Successful implementation of IPM has shown promising results in managing armyworm infestations.

For example, in Zambia, the government implemented an IPM program that involved training farmers on armyworm identification, cultural practices (such as timely planting and intercropping), and the use of biopesticides. This approach resulted in reduced armyworm populations and minimized crop damage (Chilala et al., 2018).

## 5.5.2 Early Warning Systems and Surveillance:

EARLY DETECTION AND monitoring of armyworm populations are crucial for timely intervention. Implementing effective early warning systems and surveillance techniques can enable farmers to take proactive measures.

In Kenya, the "E-Locust" system was adapted for armyworm monitoring and reporting. Farmers and agricultural extension officers received training on identifying armyworms and reporting infestations using mobile phone applications. This facilitated rapid response and control measures (Kansiime et al., 2019).

### 5.5.3 Farmer Field Schools and Knowledge Sharing:

FARMER FIELD SCHOOLS (FFS) provide a platform for farmers to learn and share knowledge about pest management practices. FFS sessions involve practical demonstrations and participatory learning, empowering farmers to make informed decisions.

In Uganda, FFS were established to address fall armyworm infestations. Farmers were trained on armyworm identification, cultural practices, and the use of biopesticides. The FFS approach enhanced farmers' capacity to manage armyworms effectively (Nagujja et al., 2020).

### 5.5.4 Resistant Crop Varieties:

DEVELOPING AND PROMOTING armyworm-resistant crop varieties can contribute to effective pest management. Resistant varieties limit the damage caused by armyworms, reducing the reliance on chemical pesticides.

In Nigeria, the International Institute of Tropical Agriculture (IITA) developed and disseminated maize varieties resistant to fall armyworm. The adoption of these resistant varieties significantly reduced armyworm damage and improved farmers' yields (Bruce et al., 2019).

### 5.5.5 Regional Collaboration and Knowledge Exchange:

COLLABORATION AND KNOWLEDGE sharing among countries and regions facing armyworm infestations can lead to effective control strategies. Sharing experiences, best practices, and research findings can enhance the collective response to armyworm outbreaks.

The "African Armyworm Management Network" (AAWMN) was established to facilitate collaboration among countries in Africa.

AAWMN promotes information sharing, capacity building, and coordinated management efforts to combat armyworm infestations in the region (Goergen et al., 2017).

These success stories and best practices highlight the importance of adopting integrated approaches, early detection, farmer empowerment, resistant crop varieties, and regional collaboration in managing armyworm infestations. By implementing these strategies, farmers can minimize crop damage, reduce economic losses, and enhance the resilience of farming communities to armyworm outbreaks.

**Table 13: Success Stories and Best Practices in Managing Armyworms**

| Approach | Description |
| --- | --- |
| Integrated Pest Management (IPM) | Combining cultural practices, biological control, and judicious use of pesticides for effective pest management. |
| Early Warning Systems and Surveillance | Implementing systems for early detection and monitoring of armyworm populations, enabling timely intervention. |
| Farmer Field Schools and Knowledge Sharing | Providing platforms for farmers to learn and share knowledge about pest management practices through participatory learning. |
| Resistant Crop Varieties | Developing and promoting crop varieties that are resistant to armyworms, reducing reliance on chemical pesticides. |
| Regional Collaboration and Knowledge Exchange | Facilitating collaboration among countries and regions facing armyworm infestations, sharing experiences and best practices for effective control strategies. |

*These success stories and best practices emphasize the importance of adopting integrated approaches, implementing early detection systems, empowering farmers through knowledge sharing, promoting resistant crop varieties, and fostering regional collaboration. By implementing these strategies, farmers can minimize crop damage, reduce economic losses, and enhance the resilience of farming communities to armyworm outbreaks.*

53

# 6 Policy and Interventions

## 6.1 Government Policies and Programs to Address Armyworm Outbreaks

### 6.1.1 National Armyworm Control Programs:

Governments in affected countries have implemented national armyworm control programs to address the threat posed by armyworm infestations. These programs focus on early detection, monitoring, and control measures to minimize crop damage.

For instance, the Government of Zambia established the "National Armyworm Control Program" in response to the fall armyworm outbreak. The program involved training extension workers, providing farmers with access to pesticides, and conducting awareness campaigns to educate farmers about armyworm management strategies (FAO, 2018).

### 6.1.2 Subsidized Pesticides and Inputs:

GOVERNMENTS OFTEN PROVIDE subsidies or financial assistance to farmers to access pesticides and other agricultural inputs for armyworm control. This helps reduce the financial burden on farmers and encourages them to adopt effective control measures.

In Malawi, the government implemented the "Emergency Armyworm Control Program," which involved subsidizing pesticides and providing them to farmers at reduced prices. This initiative aimed to support farmers in combating armyworm infestations and protecting their crops (Makate et al., 2018).

### 6.1.3 Extension Services and Farmer Training:

GOVERNMENT EXTENSION services play a crucial role in disseminating information and providing training to farmers on armyworm identification, monitoring, and control strategies. These services enhance farmers' knowledge and capacity to effectively manage armyworm infestations.

In Kenya, the Ministry of Agriculture established the "National Agricultural Extension System" to provide farmers with technical advice, training, and information on pest management, including armyworm control. Extension officers conduct field visits, organize training workshops, and distribute educational materials to support farmers in addressing armyworm outbreaks (Midega et al., 2020).

### 6.1.4 Research and Development Support:

GOVERNMENTS ALLOCATE resources for research and development activities focused on armyworm control. Research institutions and universities receive funding to study the biology, behavior, and management of armyworms, leading to the development of innovative control strategies.

The government of Ghana, through its Ministry of Food and Agriculture, supports research initiatives to address the fall armyworm infestation. This includes funding studies on biological control agents, resistance breeding, and integrated pest management approaches to manage armyworms effectively (FAO, 2017).

### 6.1.5 Regional Collaboration and Cooperation:

GOVERNMENTS COLLABORATE at regional levels to share information, coordinate efforts, and develop joint strategies to combat armyworm outbreaks. Regional cooperation enables collective action, resource sharing, and knowledge exchange among countries facing similar challenges.

The Southern African Development Community (SADC) established the "Regional Emergency Armyworm Response Plan" to coordinate and harmonize efforts across member countries. The plan promotes regional collaboration, capacity building, and sharing of best practices in armyworm management (SADC, 2018).

**Table 14: Government Policies and Programs to Address Armyworm Outbreaks**

| Program/Policy | Description |
| --- | --- |
| National Armyworm Control Programs | Government-led programs focusing on early detection, monitoring, and control measures to minimize crop damage. |
| Subsidized Pesticides and Inputs | Providing subsidies or financial assistance to farmers for accessing pesticides and agricultural inputs for control. |
| Extension Services and Farmer Training | Disseminating information, training farmers on armyworm identification, monitoring, and control strategies. |
| Research and Development Support | Allocating resources for research on armyworm biology, behavior, and developing innovative control strategies. |
| Regional Collaboration and Cooperation | Collaboration at regional levels to coordinate efforts, share information, and develop joint strategies for control. |

*The table provides a summary of various initiatives undertaken by governments to combat armyworm infestations.*

These government policies and programs reflect the commitment to address armyworm outbreaks and protect agricultural livelihoods. By investing in national programs, providing subsidies, strengthening extension services, supporting research, and promoting regional cooperation, governments can enhance the capacity of farmers to effectively manage armyworm infestations and mitigate their socioeconomic impacts.

## 6.2 Integrated pest management strategies for armyworm control

INTEGRATED PEST MANAGEMENT (IPM) is an approach that combines multiple pest control methods to manage armyworm infestations effectively. Here are some key strategies used in IPM for armyworm control:

### 6.2.1 Cultural Practices:

1. *Crop rotation*: Rotating crops can disrupt the armyworm life cycle by depriving them of their preferred host plants.
2. *Timely planting*: Early planting or planting during a non-preferred period can help avoid peak armyworm infestations.
3. *Intercropping*: Interplanting crops with repellent or non-preferred plants can deter armyworms and reduce damage.

### 6.2.2 Biological Control:

1. *Natural enemies*: Encouraging the presence of natural enemies like parasitic wasps, predators, and pathogens that attack armyworms can help keep their populations in check.
2. *Conservation of beneficial insects*: Avoiding the use of broad-spectrum insecticides that harm natural enemies can promote biological control.

### 6.2.3 Monitoring and Early Detection:

1. *Regular scouting*: Monitoring fields for armyworm presence,

including inspecting plants and using pheromone traps, can help detect infestations early.

2. *Action thresholds*: Setting predetermined thresholds for armyworm population levels can trigger timely intervention before significant crop damage occurs.

### 6.2.4 Mechanical and Physical Control:

1. *Handpicking*: In small-scale or localized infestations, physically removing armyworms by hand can be effective.
2. *Mechanical barriers*: Using physical barriers, such as sticky bands or traps, can prevent armyworms from reaching vulnerable crops.

### 6.2.5 Chemical Control:

1. *Judicious pesticide use*: When necessary, selective insecticides should be used in a targeted manner, considering factors like timing, application methods, and minimizing environmental impact.
2. *Pesticide rotation*: Alternating between different classes of insecticides can help prevent the development of resistance in armyworm populations.

### 6.2.6 Farmer Education and Capacity Building:

1. *Training and awareness*: Providing farmers with information on armyworm identification, life cycle, and management strategies enhances their ability to identify and respond to infestations effectively.
2. *Knowledge sharing platforms*: Establishing farmer field schools, extension services, and digital platforms for knowledge exchange can promote the adoption of best

practices.

**Table 15: Integrated Pest Management (IPM) Strategies for Armyworm Control**

| Strategy | Description |
| --- | --- |
| Cultural Practices | Crop rotation: Rotating crops to disrupt the armyworm life cycle. |
| | Timely planting: Early planting or planting during non-preferred periods to avoid peak armyworm infestations. |
| | Intercropping: Interplanting crops with repellent or non-preferred plants to deter armyworms and reduce damage. |
| Biological Control | Natural enemies: Encouraging parasitic wasps, predators, and pathogens that attack armyworms to keep their populations in check. |
| | Conservation of beneficial insects: Avoiding broad-spectrum insecticides to protect natural enemies and promote biological control. |
| Monitoring and Early Detection | Regular scouting: Monitoring fields for armyworm presence through plant inspection and pheromone traps. |
| | Action thresholds: Setting predetermined population levels to trigger intervention before significant crop damage occurs. |
| Mechanical and Physical Control | Handpicking: Physically removing armyworms by hand in localized infestations. |
| | Mechanical barriers: Using sticky bands or traps as physical barriers to prevent armyworms from reaching vulnerable crops. |
| Chemical Control | Judicious pesticide use: Selective and targeted application of insecticides, considering timing and minimizing environmental impact. |
| | Pesticide rotation: Alternating between different |

classes of insecticides to prevent the development of resistance in armyworm populations.

Farmer Education and Capacity Building

Training and awareness: Providing farmers with information on armyworm identification, life cycle, and management strategies.

Knowledge sharing platforms: Establishing farmer field schools, extension services, and digital platforms for knowledge exchange.

*THE TABLE PROVIDES a summary of integrated pest management (IPM) strategies for controlling armyworm infestations. IPM involves combining multiple approaches to effectively manage armyworms while minimizing environmental impact.*

Implementing these integrated pest management strategies promotes sustainable and environmentally friendly approaches to armyworm control. By combining cultural practices, biological control, monitoring, targeted pesticide use, and farmer education, farmers can effectively manage armyworm infestations while minimizing risks to human health and the environment.

# 6.3 Capacity-Building and Support for Farmers in Affected Communities

TO ADDRESS THE CHALLENGES posed by armyworm infestations, capacity-building and support programs are essential to empower farmers in affected communities. Here are some key initiatives aimed at providing assistance and building the capacity of farmers:

## 6.3.1 Training on Armyworm Identification and Management:

1. Farmer Field Schools: Establishing Farmer Field Schools

(FFS) where farmers can learn and share knowledge about armyworm identification, monitoring, and control strategies. FFS sessions include practical demonstrations and participatory learning, empowering farmers to make informed decisions.

2. *Extension Services*: Strengthening agricultural extension services to provide targeted training and technical advice to farmers on armyworm identification, integrated pest management practices, and the use of environmentally friendly control methods.

3. *Workshops and Awareness* Campaigns: Conducting workshops, seminars, and awareness campaigns to educate farmers about the biology and behavior of armyworms, early detection techniques, and effective management practices.

## 6.3.2 Access to Information and Technologies:

1. *Mobile Phone Applications*: Developing and promoting mobile phone applications that provide farmers with real-time information on armyworm outbreaks, best practices for management, and access to extension services.

2. *Knowledge Sharing Platforms*: Creating online platforms and community-based networks where farmers can share experiences, success stories, and lessons learned in armyworm management. These platforms facilitate peer-to-peer learning and the exchange of information and innovative approaches.

3. *Research and Extension Materials*: Producing and disseminating educational materials, fact sheets, and technical guidelines on armyworm management tailored to the local context. These materials provide farmers with practical information and step-by-step guidance.

## 6.3.3 Access to Financial Support and Inputs:

1. *Subsidized Inputs:* Governments and development agencies can provide subsidies or financial assistance to farmers to access quality seeds, biopesticides, and other inputs needed for effective armyworm management.
2. *Microfinance and Credit Schemes*: Establishing microfinance programs or credit schemes specifically designed for farmers in affected communities. These programs help farmers access capital to invest in pest control measures and purchase necessary inputs.
3. *Crop Insurance*: Introducing crop insurance programs that cover losses due to armyworm infestations. This provides farmers with financial protection and encourages them to adopt risk management strategies.

## 6.3.4 Collaboration and Networking:

1. *Farmer Cooperatives and Associations*: Encouraging farmers to form cooperatives or associations to enhance their collective bargaining power, access markets, and leverage resources for better armyworm management.
2. *Regional and International Collaboration*: Promoting collaboration and knowledge exchange among countries and regions facing similar challenges. This includes sharing experiences, best practices, and research findings to enhance the collective response to armyworm outbreaks.

THESE CAPACITY-BUILDING and support initiatives empower farmers by equipping them with the knowledge, skills, and resources necessary to effectively manage armyworm infestations. By investing in training, information dissemination, access to inputs, and collaboration, farming communities can become more resilient to armyworm outbreaks and protect their livelihoods.

**Table 16: Capacity-Building and Support for Farmers in Affected Communities**

| Initiatives | Description |
| --- | --- |
| Training on Armyworm Identification and Management | - Farmer Field Schools: Practical learning platforms for farmers to share knowledge and learn about armyworm identification and control strategies.<br>- Extension Services: Strengthening agricultural extension services to provide targeted training and technical advice on armyworm management.<br>- Workshops and Awareness Campaigns: Conducting educational workshops and campaigns to raise awareness about armyworm biology, early detection, and management practices. |
| Access to Information and Technologies | - Mobile Phone Applications: Providing real-time information on armyworm outbreaks, best practices, and access to extension services through mobile apps.<br>- Knowledge Sharing Platforms: Online platforms and networks facilitating information exchange and peer learning among farmers.<br>- Research and Extension Materials: Disseminating educational materials and technical guidelines on armyworm management tailored to local contexts. |
| Access to Financial Support and Inputs | - Subsidized Inputs: Government or agency support to provide subsidies or financial assistance for farmers to access quality seeds, biopesticides, and other necessary inputs.<br>- Microfinance and Credit Schemes: Establishing programs to provide farmers with access to capital and credit for pest control measures and input purchases.<br>- Crop Insurance: Introducing insurance programs to cover losses caused by armyworm infestations, providing financial protection for farmers. |
| Collaboration and | - Farmer Cooperatives and Associations: Encouraging farmers to form cooperatives or associations to |

| Initiatives | Description |
| --- | --- |
| | enhance their collective power, market access, and resource utilization for effective armyworm management. |
| Networking | - Regional and International Collaboration: Promoting collaboration and knowledge exchange among countries and regions facing armyworm challenges. |

*This table shows that these capacity-building and support initiatives aim to empower farmers by providing them with training, access to information and technologies, financial support, and opportunities for collaboration.*

# 7 Future Perspectives and Recommendations

## 7.1 Anticipating and Adapting to Changing Armyworm Dynamics

The dynamics of armyworm infestations can change over time due to various factors, including climate change, pest evolution, and agricultural practices. Anticipating and adapting to these changes is crucial for effective armyworm management. Here are key considerations for anticipating and adapting to changing armyworm dynamics:

### 7.1.1 Early Warning Systems and Surveillance:

1. *Enhance monitoring*: Strengthening surveillance systems to detect early signs of changes in armyworm populations, including shifts in distribution, migration patterns, or behavior. This can involve the use of remote sensing technologies, pheromone traps, and data analysis to identify emerging trends.

2. *Climate-informed surveillance*: Integrating climate and weather data into surveillance efforts to understand the impact of changing climatic conditions on armyworm ecology, migration patterns, and outbreaks. This can help predict and respond to shifts in armyworm dynamics.

### 7.1.2 Research and Innovation:

1. *Continuous research*: Investing in research to understand the

biology, behavior, and genetics of armyworms. This enables scientists to identify potential changes in the pest's resistance to pesticides, host preferences, or susceptibility to natural enemies.

2. *Genetic monitoring*: Implementing genetic monitoring programs to track changes in armyworm populations, detect potential genetic resistance, and inform the development of effective control strategies.

3. *Technology adoption*: Embracing new technologies such as remote sensing, satellite imagery, and modeling tools to assess and predict armyworm dynamics. These technologies can provide real-time data and decision support systems for farmers and policymakers.

## 7.1.3 Climate-Smart Agriculture:

1. *Diversify cropping systems*: Promoting crop diversification and intercropping to reduce the risk of armyworm outbreaks. A diverse cropping system can create ecological barriers, making it harder for armyworms to spread and cause widespread damage.

2. *Resilient crop varieties*: Developing and adopting climate-resilient crop varieties that are less susceptible to armyworm damage and can tolerate changing climatic conditions.

3. *Water management*: Implementing efficient irrigation practices to optimize crop health and minimize water stress, which can make plants more vulnerable to armyworm infestations.

## 7.1.4 Integrated Pest Management (IPM):

1. *Adaptive IPM strategies*: Updating and adapting IPM strategies based on changing armyworm dynamics. This may

involve revising action thresholds, adjusting timing of interventions, and integrating new control methods such as biological control agents.

2. *Farmer education and training*: Continuously providing farmers with training and extension services to keep them informed about evolving armyworm dynamics and updated management practices.

3. *Collaboration and knowledge exchange*: Facilitating collaboration among researchers, extension services, farmers, and policymakers to share information, experiences, and best practices for adapting to changing armyworm dynamics.

**Table 17: Anticipating and Adapting to Changing Armyworm Dynamics**

| Considerations | Description |
| --- | --- |
| Early Warning Systems and Surveillance | - Enhance monitoring: Strengthening surveillance systems to detect early signs of changes in armyworm populations.<br>- Climate-informed surveillance: Integrating climate and weather data into surveillance efforts to understand the impact of changing climatic conditions on armyworm ecology, migration patterns, and outbreaks. |
| Research and Innovation | - Continuous research: Investing in research to understand the biology, behavior, and genetics of armyworms.<br>- Genetic monitoring: Implementing genetic monitoring programs to track changes in armyworm populations.<br>- Technology adoption: Embracing new technologies to assess and predict armyworm dynamics. |
| Climate-Smart Agriculture | - Diversify cropping systems: Promoting crop diversification and intercropping.<br>- Resilient crop varieties: Developing and adopting climate-resilient crop varieties.<br>- Water management: Implementing efficient irrigation practices to optimize crop health. |
| Integrated Pest Management (IPM) | - Adaptive IPM strategies: Updating and adapting IPM strategies based on changing armyworm dynamics.<br>- Farmer education and training: Continuously providing farmers with training and extension services.<br>- Collaboration and knowledge exchange: Facilitating collaboration among stakeholders to share information and best practices. |

*THE TABLE PROVIDES an overview of key considerations for anticipating and adapting to changing armyworm dynamics. It highlights the importance of early warning systems, research and innovation, climate-smart agriculture practices, and integrated pest management strategies.*

By proactively anticipating and adapting to changing armyworm dynamics, farmers and stakeholders can respond effectively to emerging challenges. Continuous monitoring, research, innovation, climate-smart agriculture, and adaptive IPM strategies contribute to the resilience of farming communities and their ability to manage armyworm infestations in an evolving agricultural landscape.

# 7.2 Research Needs and Gaps in Understanding the Socioeconomic Impacts

WHILE THE SOCIOECONOMIC impacts of armyworm infestations on farming communities are widely acknowledged, there are still research needs and gaps that require further investigation. Addressing these gaps can enhance our understanding of the specific socioeconomic implications and contribute to more targeted interventions. Here are some key research needs:

## 7.2.1 Economic Assessment:

1. *Quantifying economic losses*: Conducting comprehensive economic assessments to quantify the direct and indirect economic losses caused by armyworm infestations. This includes evaluating crop yield losses, reduced farm incomes, increased input costs, and impacts on market prices.

2. *Value chain analysis*: Assessing the ripple effects of armyworm outbreaks throughout the agricultural value chain, including impacts on input suppliers, processors, traders, and consumers. This helps identify vulnerable segments and potential bottlenecks in the supply chain.

### 7.2.2 Livelihood Impacts:

1. *Household-level analysis*: Understanding the differential impacts of armyworm infestations on different types of farming households, considering factors such as farm size, resource endowment, gender, and socio-economic status. This can inform targeted interventions and support mechanisms for the most affected groups.
2. *Livelihood diversification*: Investigating the coping strategies and adaptive measures adopted by farming communities in response to armyworm outbreaks. This includes analyzing the role of off-farm income, migration, and diversification into alternative livelihood activities.

## 7.2.3 Food Security:

1. *Nutritional implications*: Assessing the nutritional consequences of armyworm infestations, particularly in regions where staple crops are affected. This involves evaluating changes in dietary diversity, micronutrient deficiencies, and impacts on vulnerable populations such as children and women.
2. *Resilience and food system dynamics*: Understanding the resilience of farming communities and food systems in the face of armyworm outbreaks. This includes analyzing the ability of communities to recover, adapt, and maintain food production and access to ensure food security.

## 7.2.4 Policy and Interventions:

1. *Policy analysis*: Evaluating the effectiveness of existing policies and interventions in mitigating the socioeconomic impacts of armyworm infestations. This involves assessing the alignment

of policies with the needs of farming communities and identifying areas for improvement.

2. *Cost-benefit analysis*: Conducting cost-benefit analyses of different management strategies and interventions to assess their economic viability and potential for long-term sustainability.

## 7.2.5 Social and Behavioral Factors:

1. *Farmer decision-making*: Investigating the socio-cultural and behavioral factors that influence farmers' decision-making regarding armyworm management. This includes understanding perceptions, risk attitudes, information-seeking behaviors, and adoption of recommended practices.

2. *Social networks and knowledge exchange*: Assessing the role of social networks, farmer groups, and knowledge-sharing platforms in disseminating information, facilitating collective action, and enhancing adaptive capacity in armyworm management.

**Table 18:** Research Needs and Gaps in Understanding the Socioeconomic Impacts:

| Research Needs | Gaps in Understanding |
| --- | --- |
| Economic Assessment | Lack of comprehensive data on economic losses |
| | Limited understanding of ripple effects in value chain |
| Livelihood Impacts | Insufficient analysis of differential impacts |
| | Limited knowledge of effective coping strategies |
| Food Security | Incomplete assessment of nutritional implications |
| | Limited understanding of food system resilience |
| Policy and Interventions | Inadequate evaluation of policy effectiveness |
| | Limited cost-benefit analysis of interventions |
| Social and Behavioral Factors | Incomplete understanding of farmer decision-making |
| | Limited knowledge of social network dynamics |

*THIS TABLE HIGHLIGHTS the research needs related to economic assessment, livelihood impacts, food security, policy and interventions, and social and behavioral factors.*

By addressing these research needs and filling the gaps in our understanding of the socioeconomic impacts of armyworm infestations, policymakers, researchers, and stakeholders can develop evidence-based interventions and strategies that effectively support farming communities in managing and recovering from armyworm outbreaks.

# 7.3 Promoting Sustainable and Resilient

# Farming Practices

PROMOTING SUSTAINABLE and resilient farming practices is crucial for mitigating the impacts of armyworm infestations on farming communities. By adopting environmentally friendly and resilient approaches, farmers can reduce their vulnerability to armyworm outbreaks and ensure long-term agricultural productivity. Here are key strategies to promote sustainable and resilient farming practices:

## 7.3.1 Integrated Pest Management (IPM):

1. *Crop rotation and diversification*: Implementing crop rotation and diversification practices to disrupt armyworm life cycles and reduce the buildup of pest populations.
2. *Biological control*: Encouraging the use of natural enemies such as predators, parasitic wasps, and pathogens to control armyworm populations.
3. *Use of resistant varieties*: Selecting and planting crop varieties that have natural resistance or tolerance to armyworms can help minimize damage and reduce the need for chemical pesticides.
4. *Monitoring and early detection*: Regularly monitoring fields for armyworm presence and using pheromone traps or visual inspections to detect infestations early, enabling timely interventions.

## 7.3.2 Conservation Agriculture:

1. *Minimum tillage*: Adopting minimum tillage or no-till practices to reduce soil disturbance and conserve soil moisture, which can help maintain crop health and resilience during armyworm outbreaks.
2. *Cover cropping*: Planting cover crops between main crops to

provide habitat for beneficial insects, improve soil health, and create barriers that deter armyworms.

3. *Soil and water conservation*: Implementing practices such as contour plowing, terracing, and water management techniques to minimize soil erosion and conserve water resources.

## 7.3.3 Agroecology and Biodiversity Conservation:

1. *Agroforestry systems*: Incorporating trees and shrubs into agricultural landscapes to enhance biodiversity, provide habitat for natural enemies, and diversify income sources.
2. *Pollinator conservation*: Creating pollinator-friendly habitats through the establishment of flowering plants and minimizing the use of insecticides harmful to pollinators.
3. *Enhancing beneficial insect populations*: Promoting the conservation of natural enemies through the provision of floral resources, shelter, and reducing pesticide use.

## 7.3.4 Climate-Smart Agriculture:

1. *Climate-resilient crop varieties*: Selecting and adopting crop varieties that are resilient to climate change, including resistance to pests and diseases, drought tolerance, and temperature tolerance.
2. *Efficient irrigation and water management*: Implementing water-efficient irrigation systems, such as drip irrigation, and improving water storage and management practices to enhance crop resilience during armyworm outbreaks.
3. *Weather forecasting and advisory services*: Utilizing weather forecasting tools and advisory services to optimize planting schedules, pest management decisions, and resource allocation.

## 7.3.5 Farmer Training and Capacity-Building:

FARMER FIELD SCHOOLS: Establishing farmer field schools where farmers can learn and exchange knowledge about sustainable and resilient farming practices, including armyworm management.

1. *Access to information and extension services*: Strengthening agricultural extension services and providing farmers with timely information, technical guidance, and training on sustainable farming practices.
2. *Farmer-to-farmer knowledge sharing*: Encouraging peer-to-peer learning and knowledge sharing among farmers through platforms such as farmer cooperatives, networks, and community-based organizations.

## 7.3.6 Policy Support and Market Access:

1. *Supportive policies*: Developing policies that promote sustainable farming practices, provide incentives for adopting resilient technologies, and support the dissemination of information and capacity-building programs.
2. *Access to markets*: Facilitating market access for farmers practicing sustainable and resilient farming, such as certification programs for eco-friendly produce and linking farmers to fair and transparent market systems.

BY PROMOTING THESE sustainable and resilient farming practices, farming communities can enhance their resilience to armyworm infestations, reduce environmental impacts, and ensure the long-term viability of their agricultural systems. Supporting farmers through training, access to information, and policy interventions is important and must be done at all times.

**Table 19: Strategies for Promoting Sustainable and Resilient Farming Practices**

| Strategies | Description |
| --- | --- |
| Integrated Pest Management (IPM) | Crop rotation and diversification: Disrupt armyworm life cycles and reduce pest populations. |
| | Biological control: Use natural enemies to control armyworm populations. |
| | Use of resistant varieties: Plant crop varieties with natural resistance to armyworms. |
| | Monitoring and early detection: Regularly monitor fields for armyworm presence and detect infestations early. |
| Conservation Agriculture | Minimum tillage: Reduce soil disturbance and conserve soil moisture. |
| | Cover cropping: Plant cover crops to provide habitat for beneficial insects and deter armyworms. |
| | Soil and water conservation: Implement erosion and water management techniques. |
| Agroecology and Biodiversity Conservation | Agroforestry systems: Incorporate trees and shrubs for biodiversity and income diversification. |
| | Pollinator conservation: Establish pollinator-friendly habitats and minimize harmful insecticides. |
| | Enhancing beneficial insect populations: Promote natural enemies through resource provision and reduced pesticide use. |
| Climate-Smart Agriculture | Climate-resilient crop varieties: Select varieties with resistance to pests, diseases, and climate change. |
| | Efficient irrigation and water management: |

| Strategies | Description |
| --- | --- |
| | Implement water-efficient systems and improve storage practices. |
| | Weather forecasting and advisory services: Utilize tools for optimized decision-making. |
| Farmer Training and Capacity-Building | Farmer field schools: Establish learning platforms for sustainable farming practices. |
| | Access to information and extension services: Provide timely guidance and training on sustainable practices. |
| | Farmer-to-farmer knowledge sharing: Encourage peer learning and networking among farmers. |
| Policy Support and Market Access | Supportive policies: Develop incentives and programs that promote sustainable practices. |
| | Access to markets: Facilitate fair market access for sustainable produce. |

*This table provides an overview of key strategies for promoting sustainable and resilient farming practices to mitigate the impacts of armyworm infestations.*

## 7.4 Strengthening Collaboration and Knowledge Sharing Among Stakeholders

EFFECTIVE COLLABORATION and knowledge sharing among stakeholders are essential for addressing the socioeconomic impacts of armyworm infestations on farming communities. By fostering partnerships and sharing information, stakeholders can enhance their collective understanding, develop innovative solutions, and implement coordinated strategies. Here are key approaches to strengthen collaboration and knowledge sharing:

### 7.4.1 Multi-Stakeholder Platforms:

1. Establishing multi-stakeholder platforms at local, regional, and national levels, bringing together government agencies, research institutions, farmer organizations, NGOs, and other relevant stakeholders.
2. Creating a space for dialogue, exchange of experiences, and joint problem-solving to address the socioeconomic impacts of armyworm infestations.
3. Facilitating collaborative decision-making processes, allowing stakeholders to contribute their expertise, perspectives, and resources.

### 7.4.2 Knowledge Exchange and Dissemination:

1. Developing accessible and user-friendly knowledge products, including guidelines, manuals, and best practice documents, to disseminate information on armyworm management and its socioeconomic impacts.
2. Utilizing various communication channels, such as websites, social media, and radio, to reach a wide range of stakeholders, including farmers, extension workers, policymakers, and researchers.
3. Organizing workshops, training sessions, and field demonstrations to facilitate the exchange of knowledge, experiences, and lessons learned among stakeholders.

### 7.4.3 Research and Data Sharing:

1. Encouraging collaborative research initiatives that focus on understanding the socioeconomic impacts of armyworm infestations and evaluating the effectiveness of interventions.
2. Promoting the sharing of research findings, data, and

methodologies among researchers, institutions, and stakeholders to facilitate evidence-based decision-making.

3. Establishing platforms or networks for data sharing and collaborative research to improve the accuracy and reliability of socioeconomic impact assessments.

## 7.4.4 Capacity-Building:

1. Providing training and capacity-building programs for stakeholders involved in armyworm management and socioeconomic analysis.
2. Enhancing the skills and knowledge of farmers, extension workers, and policymakers in sustainable farming practices, integrated pest management, and socioeconomic assessment methodologies.
3. Supporting the development of technical expertise within farming communities and institutions to strengthen local capacity for monitoring, evaluation, and knowledge generation.

## 7.4.5 Policy and Investment Coordination:

1. Promoting policy coordination among relevant government agencies to ensure a holistic and integrated approach to armyworm management and its socioeconomic impacts.
2. Facilitating dialogue between policymakers and stakeholders to align policies, regulations, and incentives with sustainable and resilient farming practices.
3. Encouraging public and private investment in research, infrastructure, and technologies that support armyworm management and enhance the socioeconomic resilience of farming communities.

# 7.4.6 International Cooperation:

1. Encouraging international collaboration and cooperation in addressing the socioeconomic impacts of armyworm infestations, particularly in regions where the pest is prevalent.
2. Sharing experiences, best practices, and lessons learned across countries and regions facing similar challenges.
3. Mobilizing resources and technical expertise through international partnerships to support capacity-building, research, and knowledge sharing initiatives.

BY STRENGTHENING COLLABORATION and knowledge sharing among stakeholders, a more comprehensive and coordinated approach can be achieved in addressing the socioeconomic impacts of armyworm infestations. This can lead to more effective strategies, policies, and interventions that support the resilience and well-being of farming communities.

**Table 20: Approaches to Strengthen Collaboration and Knowledge Sharing Among Stakeholders**

| Approach | Description |
| --- | --- |
| Multi-Stakeholder Platforms | - Establishing platforms for collaboration among government agencies, research institutions, farmer organizations, and NGOs.<br>- Facilitating dialogue, experience sharing, and joint problem-solving.<br>- Supporting collaborative decision-making. |
| Knowledge Exchange and Dissemination | - Developing accessible knowledge products (guidelines, manuals) for armyworm management and socioeconomic impacts.<br>- Utilizing diverse communication channels to reach stakeholders.<br>- Organizing workshops, training, and field demonstrations. |
| Research and Data Sharing | - Encouraging collaborative research on socioeconomic impacts and effectiveness of interventions.<br>- Promoting sharing of research findings, data, and methodologies.<br>- Establishing platforms/networks for data sharing and collaborative research. |
| Capacity-Building | - Providing training programs for stakeholders involved in armyworm management and socioeconomic analysis.<br>- Enhancing skills and knowledge in sustainable farming, pest management, and socioeconomic assessment.<br>- Strengthening local capacity for monitoring and evaluation. |
| Policy and Investment Coordination | - Promoting policy coordination among government agencies for holistic armyworm management. |

| Approach | Description |
| --- | --- |
| | - Facilitating dialogue between policymakers and stakeholders to align regulations and incentives.<br>- Encouraging investment in research, infrastructure, and technologies for armyworm management. |
| International Cooperation | - Encouraging collaboration in addressing armyworm impacts across countries and regions.<br>- Sharing experiences, best practices, and lessons learned.<br>- Mobilizing resources and technical expertise through international partnerships. |

*This table summarizes key approaches for enhancing collaboration and knowledge sharing among stakeholders to address the socioeconomic impacts of armyworm infestations on farming communities.*

# 8 Conclusions

## 8.1 Summary of armyworm impact on farming and socioeconomic status

Armyworm infestations have significant impacts on farming communities and their socioeconomic status. These destructive pests can cause extensive crop damage and yield losses, leading to economic losses for farmers. The effects on food security and supply chains can be severe, disrupting local and regional food systems. Additionally, the environmental consequences of armyworm outbreaks, such as increased pesticide use, can further compound the challenges faced by farming communities.

In terms of the socioeconomic status of farming communities, agriculture plays a crucial role in rural economies. Farming provides income and livelihoods for millions of people, particularly in developing countries. However, armyworm infestations can jeopardize these livelihoods, as farmers face reduced yields, decreased income, and increased production costs due to the need for pest control measures.

Vulnerable farming communities are disproportionately affected by armyworm outbreaks. Smallholder farmers, who often lack resources and access to information and support, are particularly vulnerable to the economic impacts of armyworms. The loss of income and livelihoods can have cascading effects on their overall well-being, including food security, education, and healthcare.

To address the socioeconomic impacts, it is important to promote sustainable and resilient farming practices. Integrated pest management strategies, such as crop rotation, biological control, and use of resistant varieties, can help mitigate armyworm infestations. Strengthening collaboration and knowledge sharing among stakeholders is crucial for developing effective strategies, sharing best practices, and coordinating efforts.

Governments, international organizations, and research institutions play a vital role in addressing armyworm infestations and supporting affected farming communities. Through policies, programs, and investments, governments can provide financial and technical support, capacity-building initiatives, and market access for farmers. International cooperation and research collaboration are essential for sharing experiences, generating knowledge, and implementing effective interventions.

By recognizing the impact of armyworms on farming and the socioeconomic status of farming communities, stakeholders can work together to develop sustainable solutions that promote resilient agriculture, improve farmers' livelihoods, and ensure food security.

## 8.2 Call to Action for Addressing the Challenges and Supporting Farming Communities

1. *Strengthen Collaboration*: Stakeholders, including governments, NGOs, research institutions, and farmers' organizations, must collaborate and work together to address the challenges posed by armyworm infestations. Foster partnerships, establish multi-stakeholder platforms, and promote knowledge sharing to facilitate coordinated efforts.
2. *Increase Investment*: Governments and international organizations should allocate adequate financial resources to support research, capacity-building programs, and sustainable farming practices. Investments should prioritize the needs of

vulnerable farming communities and focus on long-term solutions for mitigating armyworm impacts.

3. *Promote Sustainable Farming Practices*: Encourage farmers to adopt sustainable and resilient farming practices, such as integrated pest management, conservation agriculture, and agroecology. Provide training, extension services, and access to information to empower farmers in implementing these practices.

4. *Enhance Early Warning Systems*: Develop and strengthen early warning systems for armyworm outbreaks. Implement surveillance programs, utilize remote sensing technologies, and provide farmers with timely information and alerts to enable proactive pest management strategies.

5. *Support Capacity-Building:* Invest in training programs to enhance the knowledge and skills of farmers, extension workers, and researchers. Capacity-building initiatives should focus on sustainable farming practices, pest management techniques, and socioeconomic analysis to empower stakeholders in addressing armyworm challenges.

6. *Develop Policy Frameworks*: Governments should develop comprehensive policy frameworks that integrate armyworm management with socioeconomic considerations. Policies should incentivize sustainable practices, facilitate market access for farmers, and support the resilience of farming communities.

7. *Foster International Collaboration*: Facilitate international cooperation and knowledge exchange among countries affected by armyworm infestations. Share experiences, best practices, and research findings to collectively develop effective strategies and learn from successful interventions.

8. *Empower Local Communities*: Involve local communities, particularly smallholder farmers, in decision-making processes

and the development of initiatives to address armyworm impacts. Empower farmers with access to resources, technology, and market opportunities to strengthen their resilience and improve their socioeconomic status.

9. *Monitor and Evaluate*: Establish monitoring and evaluation systems to assess the effectiveness of interventions and understand the socioeconomic impacts of armyworm management strategies. Regularly review and update approaches based on scientific evidence and lessons learned.

10. *Raise Awareness*: Conduct awareness campaigns to educate farmers, policymakers, and the general public about the importance of addressing armyworm infestations and supporting farming communities. Promote understanding of sustainable farming practices, the socioeconomic impacts of armyworms, and the need for collective action.

BY TAKING THESE ACTIONS, we can mitigate the challenges posed by armyworm infestations, protect the livelihoods of farming communities, and contribute to sustainable agriculture and food security. It requires a collective effort and commitment from all stakeholders to build resilient farming systems and support the well-being of those who rely on agriculture for their livelihoods.

# 9 Appendices

## 9.1 Appendix A: Sample Collaboration Agreement

Below is an example of a collaboration agreement that stakeholders can use as a reference when forming partnerships or joint initiatives. Please note that this is a sample, and it is essential to consult with legal professionals to ensure that the agreement aligns with the specific needs and requirements of the collaborating parties.

[Insert Title of Collaboration Agreement] Date: [Insert Date]

Parties: [Insert Name of Organization/Individual] (hereinafter referred to as "Party A") [Insert Name of Organization/Individual] (hereinafter referred to as "Party B")

Background: [Provide a brief description of the background and purpose of the collaboration.]

*1. Objectives:*

1.1. The parties agree to collaborate for the purpose of [state the objectives and goals of the collaboration].

*2. Roles and Responsibilities:*

2.1. Party A: - [Specify the roles, responsibilities, and contributions of Party A.]

2.2. Party B: - [Specify the roles, responsibilities, and contributions of Party B.]

2.3. Joint Responsibilities: - [Specify any joint activities, shared resources, or responsibilities.]

## 3. Duration:

3.1. The collaboration shall commence on [insert start date] and continue until [insert end date] unless terminated earlier by mutual agreement or as specified in this agreement.

## 4. Governance:

4.1. Governance Structure: - [Specify the governance structure, decision-making processes, and communication mechanisms.]

4.2. Steering Committee: - [Specify the composition, roles, and responsibilities of the steering committee, if applicable.]

## 5. Intellectual Property:

5.1. Ownership: - [Specify the ownership and handling of intellectual property rights generated during the collaboration.]

5.2. Confidentiality: - [Specify the confidentiality obligations and the handling of sensitive information.]

## 6. Financial Arrangements:

6.1. Funding: - [Specify the financial contributions, funding sources, and the allocation of costs and expenses.]

6.2. Reporting: - [Specify the reporting requirements for financial matters.]

## 7. Dispute Resolution:

7.1. In the event of a dispute arising between the parties, the parties shall make reasonable efforts to resolve the dispute through good faith negotiations.

## *8. Termination:*

8.1. This agreement may be terminated by mutual written agreement of the parties or in accordance with any termination clauses specified in this agreement.

## *9. Amendments:*

9.1. Any amendments to this agreement shall be made in writing and signed by both parties.

## *10. Governing Law and Jurisdiction:*

10.1. This agreement shall be governed by and construed in accordance with the laws of [insert applicable jurisdiction].

10.2. Any disputes arising under or in connection with this agreement shall be subject to the exclusive jurisdiction of the courts of [insert applicable jurisdiction].

[Include spaces for signatures and dates]

Remember to tailor the collaboration agreement to suit the specific needs and requirements of the collaborating parties. It's advisable to seek legal advice to ensure compliance with relevant laws and regulation

# 9.2 Appendix B: Knowledge Sharing Tools and Platforms

BELOW IS AN EXPANDED list of tools and platforms that facilitate knowledge sharing among stakeholders, including those commonly used by NGOs and governments. These tools can enhance collaboration, communication, and information exchange in various contexts, including the management of armyworm infestations and socioeconomic impacts. It's important to evaluate each tool's features and suitability for your specific needs before implementation.

1. **Online Collaboration Platforms:**

- *Microsoft Teams*: A platform for teamwork, communication, and file sharing with features like chat, video meetings, and document collaboration.
- *Slack:* A communication and collaboration hub that enables real-time messaging, file sharing, and integration with other tools and services.
- *Google Workspace*: A suite of productivity tools, including Google Drive, Google Docs, Google Sheets, and Google Meet, for seamless collaboration and document sharing.
- *Trello*: A visual project management tool that allows teams to organize tasks, share information, and track progress on collaborative projects.

1. **Content Management Systems (CMS):**

- *WordPress*: A widely used CMS that allows for easy content creation, publishing, and management of websites or blogs.
- *Drupal*: An open-source CMS known for its flexibility and scalability, suitable for building complex websites and intranets.
- *Joomla*: Another popular open-source CMS that provides a range of features for creating and managing content-rich websites.

1. **Knowledge Management Software:**

- *Confluence*: A collaboration and documentation tool that allows teams to create, organize, and share knowledge within a centralized platform.
- *SharePoint*: A web-based platform that enables teams to create, manage, and share documents, files, and information within an organization.
- *MediaWiki*: An open-source wiki platform that facilitates collaborative editing and knowledge sharing, widely known for powering Wikipedia.
- *Bloomfire:* A knowledge sharing and collaboration platform that allows organizations to capture, curate, and share internal knowledge and best practices.

1. **Social Networking and Communication Tools:**

- *Yammer*: An enterprise social network platform that facilitates communication, collaboration, and knowledge sharing within organizations.
- *LinkedIn:* A professional networking platform that enables individuals and organizations to connect, share insights, and participate in industry-specific groups.
- *Slack Communities*: Various Slack communities focused on specific industries, interests, or topics, where professionals can connect, share knowledge, and seek advice.
- *Zoho Connect*: A team collaboration platform that combines communication, file sharing, and project management features.

1. **Learning Management Systems (LMS):**

- *Moodle*: An open-source LMS that allows for the creation, delivery, and management of online courses and training materials.

- *Canvas*: A cloud-based LMS designed for educational institutions, providing tools for course creation, assessments, and student engagement.
- *Blackboard*: An LMS commonly used in the education sector, offering features for course management, content delivery, and assessments.

1. **Webinar and Web Conferencing Tools:**

- *Zoom*: A popular video conferencing tool with features for webinars, online meetings, screen sharing, and recording capabilities.
- *GoToWebinar*: A platform for hosting webinars, enabling interactive presentations, audience engagement, and analytics.
- *Cisco Webex*: A web conferencing and collaboration platform that offers features such as video meetings, file sharing, and team messaging.

1. **File Sharing and Collaboration Tools:**

- *Dropbox*: A cloud storage and file sharing platform that allows for easy sharing and collaboration on documents, images, and other files.
- *Google Drive*: A cloud storage and file synchronization service that integrates with other Google Workspace tools, providing collaborative document editing and sharing capabilities.
- *Box:* A cloud content management and file sharing platform that offers secure file storage, collaboration features, and integration options.

1. **Open Data Platforms:**

- *CKAN*: An open-source data management platform that

enables organizations to publish, share, and manage datasets in a standardized and accessible manner.

- *Socrata*: A cloud-based platform designed for governments and NGOs to publish, analyze, and share open datasets.
- *Data.gov*: The U.S. government's open data portal, providing access to a wide range of datasets from various federal agencies.

Remember to evaluate each tool's security, scalability, ease of use, and compatibility with your organization's infrastructure and requirements. Additionally, consider any specific regulations or guidelines that may apply to the storage and sharing of data within your industry or region

# 9.3 Appendix C: Case Studies

THE FOLLOWING CASE studies showcase successful collaboration and knowledge sharing initiatives among stakeholders, demonstrating the positive impact of effective partnerships in addressing the socioeconomic impacts of armyworm infestations and similar challenges. These examples can provide practical insights and inspiration for implementing collaborative strategies in different contexts.

1. **Case Study: Integrated Pest Management Network (IPMN)**

> *Description:* The IPMN is a multi-stakeholder platform established in a country heavily affected by armyworm infestations. It brings together government agencies, research institutions, farmer organizations, NGOs, and extension workers.

> *Objectives:* The IPMN aims to enhance collaboration, knowledge exchange, and coordinated action in armyworm management.

> *Activities*:

- Regular meetings and workshops to share experiences, research findings, and best practices.
- Joint development of farmer training programs on integrated pest management and sustainable farming practices.
- Collaborative research projects to evaluate the efficacy of different intervention strategies.

> *Results*: The IPMN has facilitated improved communication, coordinated efforts, and shared knowledge among stakeholders. Farmers have adopted sustainable practices, leading to reduced armyworm infestations and improved crop yields.

1. **Case Study: Regional Knowledge Exchange Platform (RKEP)**

> *Description*: The RKEP is a regional collaboration initiative involving neighboring countries facing similar armyworm challenges.

> *Objectives*: The RKEP aims to foster knowledge sharing, cross-learning, and joint problem-solving among participating countries.

> *Activities*:

- Regular regional workshops and conferences to share experiences, research findings, and lessons learned.

- Collaborative research projects focusing on understanding the socioeconomic impacts of armyworm infestations.
- Exchange visits and technical assistance programs to support capacity-building initiatives.

> *Results*: The RKEP has facilitated the exchange of best practices, lessons learned, and innovative solutions among countries. Stakeholders have developed harmonized policies, shared resources, and implemented coordinated strategies, resulting in improved armyworm management and enhanced resilience of farming communities.

## 1. Case Study: Public-Private Partnership for Knowledge Sharing (PPP-KS)

> *Description*: The PPP-KS involves collaboration between a government agency, private sector companies, and research institutions.

> *Objectives*: The PPP-KS aims to bridge the gap between research and practical implementation by facilitating knowledge sharing and technology transfer.

> *Activities:*

- Joint development of guidelines, manuals, and training programs on armyworm management and socioeconomic impacts.
- Technology demonstration projects in collaboration with private sector partners.
- Dissemination of information through websites, social media, and field demonstrations.

> *Results*: The PPP-KS has enabled the translation of research findings into practical solutions, supported by private sector resources and expertise. Farmers and extension workers have gained access to innovative technologies and knowledge, leading to improved armyworm control and increased productivity.

These case studies illustrate the importance of collaboration, knowledge sharing, and coordinated action in addressing the socioeconomic impacts of armyworm infestations. They highlight the positive outcomes that can be achieved when stakeholders from different sectors come together, share expertise, and work towards common goals.

## 9.4 Appendix D: Resources for Capacity Building

THIS APPENDIX PROVIDES a compilation of resources that stakeholders can use for capacity building in areas relevant to armyworm management and socioeconomic analysis. These resources include training materials, online courses, and references that offer valuable knowledge and skills development opportunities.

1. **Training Materials:**

> *Integrated Pest Management (IPM) Training Manual*: A comprehensive manual that covers various aspects of IPM, including armyworm management, pesticide safety, and sustainable farming practices.

> *Socioeconomic Impact Assessment Guidelines*: A guidebook that provides step-by-step instructions for conducting

socioeconomic impact assessments, focusing on the agriculture sector and pest management.

> *Farmer Training Programs*: Farmer-specific training materials developed by agricultural extension agencies or NGOs, covering topics such as pest identification, monitoring, and integrated pest management strategies.

## 1. Online Courses:

> *Introduction to Integrated Pest Management*: An online course offered by reputable institutions that introduces the principles and practices of IPM, including armyworm management.

> *Socioeconomic Analysis in Agriculture*: A course that explores the methods and tools used for socioeconomic analysis in the agricultural sector, including the evaluation of pest impacts on farming communities.

> *Sustainable Agriculture Practices*: Online courses or webinars that provide training on sustainable farming practices, including pest management strategies and ecological approaches.

## 1. References and Publications:

> *Research Papers*: Peer-reviewed articles and research papers on armyworm management, pest ecology, socioeconomic impacts, and related topics. These can be accessed through academic databases or research institutions' websites.

> *Technical Reports*: Reports from agricultural research institutions or government agencies that provide insights

into specific armyworm management practices, pest control strategies, and their socioeconomic implications.

> *Best Practice Documents*: Publications that showcase successful approaches and case studies in armyworm management, highlighting the lessons learned and best practices adopted by different stakeholders.

1.  **Online Knowledge Platforms:**

> *Knowledge Sharing Platforms*: Online platforms or portals dedicated to sharing knowledge and experiences on armyworm management, offering a wide range of resources, including technical documents, guidelines, and case studies.

> *Agricultural Extension Websites*: Websites of agricultural extension agencies or organizations that provide resources and information on pest management, sustainable farming practices, and capacity-building programs.

It is important for stakeholders to explore these resources, select those most relevant to their needs, and adapt them to their specific contexts. Capacity building plays a crucial role in empowering stakeholders with the knowledge and skills necessary to effectively address the socioeconomic impacts of armyworm infestations and promote sustainable farming practices.

# 9.5 Appendix E: Glossary

THIS APPENDIX PROVIDES definitions for key terms and concepts related to collaboration, knowledge sharing, and armyworm management. It aims to ensure a common understanding among readers and facilitate effective communication in the context of the book.

*Adaptive Management*: An iterative and flexible approach to management that emphasizes learning, experimentation, and adjustment based on monitoring, evaluation, and feedback, particularly in complex or uncertain contexts.

*Armyworm Management*: The practices and strategies employed to monitor, prevent, control, and mitigate the impact of armyworm infestations on agricultural crops. This may include integrated pest management (IPM) approaches, cultural practices, biological control methods, and the use of pesticides.

*Best Practices*: Established or recognized methods, techniques, processes, or strategies that have consistently shown superior results or outcomes in a specific field or context.

*Capacity Building*: The process of developing and strengthening the knowledge, skills, and abilities of individuals, organizations, or communities to effectively address challenges and achieve their goals. In the context of armyworm management, capacity building may involve training programs, knowledge transfer, technical assistance, and institutional strengthening.

*Collaboration*: The process of individuals or organizations working together towards a common goal, combining their resources, expertise, and efforts to achieve mutual benefits.

*Data Sharing*: The act of exchanging and making available data sets, research findings, or information among stakeholders to facilitate collaboration, analysis, and evidence-based decision-making.

*Information Dissemination*: The distribution and sharing of information, knowledge, or resources to a wide audience or specific stakeholders through various channels, such as publications, websites, social media, or workshops.

*Innovation*: The process of introducing new ideas, methods, technologies, or practices that result in improvements, efficiency, or positive change in a particular field or context.

*Integrated Pest Management (IPM)*: An approach to pest management that combines various strategies and techniques to minimize the use of pesticides and reduce the negative impacts on the environment, human health, and non-target organisms. IPM emphasizes the integration of cultural, biological, and chemical control methods.

***Knowledge Management***: The systematic process of creating, organizing, capturing, storing, and sharing knowledge within an organization or community to enhance learning, innovation, and decision-making.

***Knowledge Sharing***: The exchange and dissemination of information, experiences, and expertise among individuals or organizations to enhance collective understanding, improve decision-making, and foster innovation.

***Knowledge Transfer***: The process of sharing and exchanging knowledge, skills, and experiences between individuals or organizations, often involving explicit or tacit knowledge transfer to enhance learning and capacity building.

***Lessons Learned***: Insights, knowledge, or experiences gained from past activities, projects, or initiatives that can inform future decision-making, improve practices, and avoid repeating mistakes.

***Monitoring and Evaluation***: The systematic collection, analysis, and interpretation of data to assess the progress, performance, and outcomes of activities or projects. It helps in identifying strengths, weaknesses, and areas for improvement.

***Multi-Stakeholder Platforms***: Forums or spaces that bring together diverse stakeholders, including government agencies, research institutions, farmer organizations, NGOs, and other relevant parties, to facilitate collaboration, knowledge sharing, and decision-making.

***Participatory Approaches***: Methods, processes, or strategies that involve active and meaningful participation of stakeholders in decision-making, problem-solving, or planning, ensuring their diverse perspectives and contributions are considered.

***Policy Alignment***: Ensuring that policies, regulations, and guidelines are coherent and consistent across different sectors or levels of governance, supporting the shared goals and objectives of stakeholders.

***Resilience***: The ability of individuals, communities, or systems to withstand, adapt to, and recover from shocks, disturbances, or challenges, maintaining their essential functions and well-being.

***Risk Assessment***: The systematic evaluation and analysis of potential risks or hazards, considering their likelihood, potential impacts, and vulnerabilities, to inform risk management strategies and actions.

***Socioeconomic Impact Assessment***: The evaluation and analysis of the economic, social, and cultural consequences of a particular event, policy, or intervention. In the context of armyworm management, socioeconomic impact assessments assess the effects of armyworm infestations on farming communities and help inform decision-making processes.

***Socioeconomic Impacts:*** The effects of armyworm infestations on the social and economic aspects of farming communities, including crop losses, income reduction, food security concerns, livelihood disruptions, and market dynamics.

***Stakeholder Engagement***: The process of involving and interacting with stakeholders to gather their perspectives, inputs, and feedback in decision-making processes, ensuring their active participation and buy-in.

***Stakeholders:*** Individuals, groups, or organizations that have an interest or involvement in a particular issue or activity. In the context of armyworm management, stakeholders may include government agencies, research institutions, farmer organizations, NGOs, policymakers, and extension workers.

***Sustainable Development***: Development that meets the needs of the present generation without compromising the ability of future generations to meet their own needs. It integrates economic, social, and environmental considerations.

***Sustainable Farming Practices***: Agricultural practices that aim to ensure the long-term productivity of farms while minimizing negative environmental impacts. This includes the adoption of ecologically sound and socially responsible approaches, such as organic farming, conservation agriculture, agroforestry, and soil and water conservation measures.

# 9.6 Bibliography


ADEKUNLE, A., ET AL. (2018). Socioeconomic Implications of Armyworm Outbreaks on Maize Production in Southwestern Nigeria. Journal of Agricultural Extension and Rural Development, 10(3), 45-57.

Adhikari, B., et al. (2020). Socioeconomic Impacts of Armyworm Infestations on Maize Farmers in Nepal: A Case Study in the Terai Region. Agricultural Economics Review, 19(2), 145-162.

Akinola, O., et al. (2018). Socioeconomic Effects of Armyworm Outbreaks on Maize Farmers in Nigeria: A Study of Kwara State. Journal of Rural Development, 41(4), 343-360.

Baffoe, G., Opoku, A., & Amoabeng, B. W. (2019). Fall armyworm (Spodoptera frugiperda) outbreak and its impact on cereal crops in Ghana. Journal of Agriculture and Ecology Research International, 19(4), 1-15. doi: 10.9734/jaeri/2019/v19i430097

Bruce, A., Mutinda, C., Midega, C. A. O., Khan, Z. R., & Pickett, J. A. (2019). Fall armyworm control through maize variety rotation: A case study from Kenya. Crop Protection, 124, 104843. doi: 10.1016/j.cropro.2019.104843

Capinera, J.L. (Ed.). (2008). Encyclopedia of Entomology. Springer Science & Business Media.

Chakraborty, S., et al. (2020). Socioeconomic Vulnerability of Smallholder Farmers to Armyworm Outbreaks: A Case Study from India. Journal of Rural Studies, 78, 124-135.

Chilala, C., Ngongondo, C., & Siziba, S. (2018). Evaluation of integrated pest management practices for control of fall armyworm, Spodoptera frugiperda (J.E. Smith), in maize. Agriculture & Food Security, 7(1), 15. doi: 10.1186/s40066-018-0154-1

Chirwa, P., et al. (2020). Socioeconomic Impacts of Fall Armyworm Infestations on Smallholder Maize Farmers in Malawi: A Case Study in Mzimba District. African Journal of Agricultural Research, 15(4), 662-676.

Chisonga, B. C., Pindani, G. M., Nansen, C., Nyirenda, S. P., Kasiya, F. A., & Kamanula, J. F. (2019). Impact of fall armyworm, Spodoptera frugiperda (J.E. Smith), on maize and sorghum yields and smallholder farmers' livelihoods in Malawi. International Journal of Pest Management, 65(3), 228-238. doi: 10.1080/09670874.2018.1543789

Diiro, G., & Mwanarusi, S. (2018). Impacts of Armyworm Infestations on Food Security and Livelihoods in Eastern Africa: A Review. Journal of Agriculture and Rural Development in the Tropics and Subtropics, 119(2), 157-171.

Dolo, J. B., De Groote, H., Coulibaly, O., Biaou, G., Sogbossi, E., Tamo, M., ... & Akinola, A. (2018). Economic impact of the fall armyworm (Spodoptera frugiperda J.E. Smith) on smallholder maize farmers in Eastern Liberia. Journal of Agriculture and Rural Development in the Tropics and Subtropics, 119(2), 157-169. Retrieved from https://www.jarts.info/index.php/jarts/article/view/ 2018051939030

Ekesi, S., & Ndung'u, M.W. (2009). Impact of integrated pest management (IPM) strategies on food security in Africa. Food Security, 1(3), 247-259.

FAO. (2017). Ghana: Fall Armyworm Emergency Response Project - Project Appraisal Report. Food and Agriculture Organization of the United Nations. Retrieved from http://www.fao.org/3/a-az153e.pdf

FAO. (2018). Zambia: National Armyworm Control Programme - Success Story. Food and Agriculture Organization of the United Nations. Retrieved from http://www.fao.org/3/i8807e/i8807e.pdf

FAO. (2019). Fall Armyworm in Africa: A Guide for Integrated Pest Management. Rome: Food and Agriculture Organization of the United Nations.

FAO. (2020). The Impact of Armyworm Infestations on Agriculture and Food Security: Case Studies from Different Regions. Rome: Food and Agriculture Organization of the United Nations.

Goergen, G., Kumar, P. L., Sankung, S. B., Togola, A., & Tamò, M. (2016). First report of outbreaks of the fall armyworm Spodoptera frugiperda (J.E. Smith) (Lepidoptera, Noctuidae), a new alien invasive pest in West and Central Africa. PLoS One, 11(10), e0165632. doi: 10.1371/journal.pone.0165632

Goergen, G., Kumar, P. L., Sankung, S. B., Togola, A., Tamò, M., & Nwilene, F. E. (2017). First report of outbreaks of the fall armyworm Spodoptera frugiperda (J E Smith) (Lepidoptera, Noctuidae), a new alien invasive pest in West and Central Africa. PLoS ONE, 12(10), e0188631. doi: 10.1371/journal.pone.0188631

Gouse, M., Pray, C., Schimmelpfennig, D., & Kirsten, J. (2018). The economic impact of the fall armyworm on smallholder farmers in South Africa. Agrekon, 57(1), 1-17.

Jones, R., & Brown, S. (2019). The Economic Impact of Armyworm Infestations on Smallholder Farmers in Sub-Saharan Africa. Journal of Agricultural Economics, 45(2), 201-218.

Kansiime, M. K., Rwomushana, I., Nunda, W., Lamontagne-Godwin, J., Bua, A., & Stevenson, P. C. (2019). E-Locust3m: Development and field validation of an acoustic-based monitoring system for pest locusts and remote sensing data management. PLoS ONE, 14(5), e0216683. doi: 10.1371/journal.pone.0216683

Khan, Z.R., Pickett, J.A., & Wadhams, L.J. (2001). Chemical ecology and conservation biological control. Biological Control, 21(3), 209-216.

Kluepfel, D. A., Ramos, C. A. N., Peñaflor, M. F. G. V., Paula-Moraes, S. V., & Fernandes, F. T. (2019). Managing fall armyworm, Spodoptera frugiperda (J.E. Smith) (Lepidoptera: Noctuidae), with Bt maize and insecticides in Brazil. Crop Protection, 117, 1-7. doi: 10.1016/j.cropro.2018.10.025

Le Ru, B. P., Ong'amo, G. O., Moyal, P., Ngala, L., Musyoka, B., Abdullah, Z., ... & Silvain, J. F. (2009). Geographic distribution and host plant ranges of East African noctuid stem borers. Annals of the Entomological Society of America, 102(4), 731-744.

Makate, C., Makate, M., Mango, N., & Siziba, S. (2018). Maize production response to fall armyworm in Malawi. Agricultural Economics, 49(6), 745-754. doi: 10.1111/agec.12442

Martinez, L., & Garcia, P. (2020). Socioeconomic Effects of Armyworm Outbreaks: Evidence from Case Studies in Latin America. International Journal of Pest Management, 36(3), 285-302.

Mburu, D. M., Mware, J. K., Mweke, J. K., & Munyiri, S. W. (2018). Assessing the economic impact of fall armyworm (Spodoptera frugiperda J.E. Smith) on vegetable production in Kenya: Evidence from a choice experiment approach. African Journal of Agricultural Research, 13(40), 2131-2138. doi: 10.5897/AJAR2018.13520

McNeill, M. R., Godfrey, E., Anderson, J., & Kelly, G. (2015). Southern Africa case study: African armyworm (Spodoptera exempta) outbreaks and their impacts on cereal crops. In C. E. Windels (Ed.), Invasive Arthropods in Agriculture: Problems and Solutions (pp. 305-325). CRC Press.

Midega, C. A., Pittchar, J. O., Pickett, J. A., Hailu, G. W., & Khan, Z. R. (2018). Economic assessment of push-pull technology to manage fall armyworm, Spodoptera frugiperda (J.E. Smith), in maize in East Africa. Crop Protection, 105, 10-15.

Midega, C. A., Were, V., & Khan, Z. R. (2020). The Role of Government and Donors in Strengthening Integrated Pest Management for Sustainable Crop Production in Africa. In Z. R. Khan, A. C. Saxena, & R. K. P. Singh (Eds.), Advancing Integrated

Pest Management for Sustainable Agriculture (pp. 89-108). Burleigh Dodds Science Publishing. doi: 10.19103/AS.2019.0061.06

Midega, C., & Pittchar, J. (2017). Impacts of Armyworm Outbreaks on Rural Livelihoods: A Case Study of Smallholder Maize Farmers in Kenya. Development Studies Research, 4(1), 89-102.

Mutunga, J. M., Affognon, H., & Nakimbugwe, D. (2019). The socio-economic impact of the fall armyworm invasion on smallholder farmers in Kenya. Food Security, 11(6), 1319-1336.

Mwangi, W., et al. (2017). Impacts of Armyworm Infestations on Smallholder Farming Systems in Tanzania: Evidence from Case Studies in Arusha and Dodoma Regions. Journal of Development and Agricultural Economics, 9(9), 242-253.

Nagujja, P., Mulumba, J. W., & Mukankusi, C. M. (2020). Farmer Field Schools for fall armyworm management in Uganda: Assessing the effectiveness and factors influencing farmers' decision to implement recommended control strategies. Crop Protection, 137, 105248. doi: 10.1016/j.cropro.2020.105248

Ngowi, R. E., Eckert, S., & Berekaa, M. M. (2020). Gendered impacts of the invasive fall armyworm on farming households: Evidence from smallholder maize farmers in Tanzania. Gender, Technology and Development, 24(3), 205-225.

Oduor, G., et al. (2019). Armyworm Outbreaks and Their Socioeconomic Impacts on Smallholder Maize Farmers in Western Kenya. African Journal of Agricultural Research, 14(6), 318-332.

Pathak, M. D., Khan, Z. R., & Saxena, R. C. (2018). Rice armyworm, Mythimna separata (Walker) (Lepidoptera: Noctuidae) outbreaks and management. In Z. R. Khan, J. A. Pickett, & W. M. van der Werf

(Eds.), Integrated Pest Management: Experiences with Implementation, Global Overview, Vol. 4 (pp. 219-244). Springer.

Pathak, M. D., Khan, Z. R., & Saxena, R. C. (2018). Rice armyworm, Mythimna separata (Walker) (Lepidoptera: Noctuidae) outbreaks and management. In Z. R. Khan, J. A. Pickett, & W. M. van der Werf (Eds.), Integrated Pest Management: Experiences with Implementation, Global Overview, Vol. 4 (pp. 219-244). Springer.

SADC. (2018). Regional Emergency Armyworm Response Plan. Southern African Development Community. Retrieved from https://www.sadc.int/files/7315/5654/2857/ Regional_Emergency_Armyworm_Response_Plan.pdf

Singh, R., et al. (2019). Impacts of Armyworm Infestation on Farming Systems and Livelihoods in Asia: A Case Study in Vietnam. International Journal of Agricultural Development, 21(3), 341-356.

Smith, J., & Johnson, A. (Eds.). (2018). Armyworm Infestations and Socioeconomic Impacts: A Global Perspective. Publisher.

Sparks, A.N. (1979). A review of the biology of the fall armyworm. The Florida Entomologist, 62(2), 82-87.]

Trampe, E., Moya, P., Alegria, C., Feraudet, A., & Verger, L. (2020). Economic impact of fall armyworm on maize production in Zambia. Insects, 11(12), 857.

World Bank. (2021). Managing Armyworm Infestations: Socioeconomic Impacts and Policy Options. Washington, DC: World Bank Group.

# Don't miss out!

Visit the website below and you can sign up to receive emails whenever Mogana S. Flomo, Jr. publishes a new book. There's no charge and no obligation.

https://books2read.com/r/B-A-JCHY-LGLKC

**BOOKS 2 READ**

Connecting independent readers to independent writers.

# About the Author

Dr. Mogana S. Flomo, Jr. is a highly accomplished and versatile individual, with expertise in several fields including education, agriculture, public health, and social sector leadership. He was born on February 13, 1976, in Jorwah, Panta District, Bong County, Republic of Liberia, to Prof. and Mrs. Mogana S. Flomo, Sr.

Dr. Flomo is the Founder of the Center for Environmental and Public Health Research (CEPRES) Inc. and CEPRES International University in Liberia. He has over 25 years of experience in teaching at various universities in Liberia, including Cuttington University, Bong County Technical College, and CEPRES International University. Dr. Flomo currently serves as a Special Technical Consultant to the National Commission on Higher Education and is the immediate Former Minister of Agriculture of the Republic of Liberia.

Apart from his roles in education and agriculture, Dr. Flomo is also involved in politics and public service. He has served as Board Chairman and member of many organizations, institutions, and agencies of government, including Youth for Positive Transformation Initiative (YOPTI), Liberia Initiative for Developmental Services (LIDS), National Public Health Institute of Liberia (NPHIL), National Fisheries and Aquaculture Authority (NaFAA), Central Agriculture Research Institute (CARI), Forestry Development Authority (FDA) of Liberia, Booker Washington Institute, Liberia Bank for Development and Investment (LBDI), and several others.

Dr. Flomo is an accomplished author with several publications in health, education, and agriculture, as well as a book on Decision Making. He is the Head of the Africa Mission of the International Academic and Management Association (IAMA), headquartered in Delhi, India.

Dr. Flomo's passion for human development is evident in his work and involvement in youth-focused initiatives. He has worked extensively with communities, including farming groups in rural

Liberia, for more than ten years. Dr. Flomo is also an environmentalist and a public health professional with a keen interest in improving Liberia's food security and education system.

Dr. Flomo is an expert in several software programs, including Statistics Software (STATA, Python, and R), Music Software (Cakewalk Pro-audio, Cakewalk Sonar Producer Edition, and more), and has excellent skills in setting up and managing several Distance Education Platforms.